Self-Assembly of Materials and Their Applications

Edited by Hemali Rathnayake, Gayani Pathiraja and Eram Sharmin

Published in London, United Kingdom

Self-Assembly of Materials and Their Applications
http://dx.doi.org/10.5772/intechopen.102269
Edited by Hemali Rathnayake, Gayani Pathiraja and Eram Sharmin

Contributors
Marwa Ben Chobba, Maduka L. Weththimuni, Mouna Messaoud, Clara Urzi, Ramzi Maalej, Maurizio Licchelli, Yong Zhong, Shiqi Liu, Zengyin Li, Feng Bai, Huiwei Tong, Beheshteh Sohrabi, Ingrid Ponce, Ana María Méndez-Torres, José H Zagal, Sakina Mussarat, Attique Ur Rehman Khan, Muhammad Adnan, Ruqia Nazir, Shaheen Begum, Hemali Rathnayake

First published in London, United Kingdom, 2023 by IntechOpen
IntechOpen is the global imprint of INTECHOPEN LIMITED, registered in England and Wales, registration number: 11086078, 5 Princes Gate Court, London, SW7 2QJ, United Kingdom

British Library Cataloguing-in-Publication Data
A catalogue record for this book is available from the British Library

Additional hard and PDF copies can be obtained from orders@intechopen.com

Self-Assembly of Materials and Their Applications
Edited by Hemali Rathnayake, Gayani Pathiraja and Eram Sharmin
p. cm.
Print ISBN 978-1-83969-701-2
Online ISBN 978-1-83969-702-9
eBook (PDF) ISBN 978-1-83969-703-6

Meet the editors

Dr. Hemali Rathnayake is an associate professor in the Department of Nanoscience, Joint School of Nanoscience and Nanoengineering (JSNN), University of North Carolina at Greensboro (UNCG), USA. She obtained her BS in Chemistry from the University of Peradeniya, Sri Lanka, in 2000 and her Ph.D. from the Department of Chemistry, University of Massachusetts Amherst (UMass), USA, in 2007. Dr. Rathnayake is a pioneer scientist and a chemist in nanomaterials chemistry with a focus on the interfacial interaction of nanomaterials, molecules, macromolecules, and polymers in homogeneous and heterogeneous media. Her research on the design, synthesis, self-assembly, and application of well-defined superstructures in nanoelectronics, environmental remediation, and sustainable energy has impacted the scientific community with highly rated peer-reviewed journal publications, patents, and more than eighty invited talks to scientific and non-scientific communities.

Dr. Gayani Pathiraja is a postdoctoral research scholar in the Department of Nanoscience, Joint School of Nanoscience and Nanoengineering (JSNN), University of North Carolina at Greensboro (UNCG), USA. She earned her bachelor's degree in Materials Science and Technology from Uva Wellassa University, Sri Lanka, and her master's degree in Electrochemistry/Environmental Engineering from the University of Peradeniya, Sri Lanka. She obtained her Ph.D. in Nanoscience from UNCG in 2021 with her investigation of the crystal growth mechanism and kinetics of ultrathin copper hydroxide/oxide nanowire self-assembly. Dr. Pathiraja began her academic career as a lecturer in the Department of Engineering Technology, University of Ruhuna, Sri Lanka in 2016. She is a co-author of several peer-reviewed journal publications and a book chapter, and she has presented her work at several regional, international, and national conferences.

Dr. Eram Sharmin is an associate professor in the Department of Pharmaceutical Chemistry, College of Pharmacy, Umm Al-Qura University, Saudi Arabia. She received a BSc in Chemistry and an MSc in Organic Chemistry from Aligarh Muslim University (AMU), India in 1998 and 2000, respectively. She obtained her Ph.D. in Chemistry from Jamia Millia Islamia (JMI) - A Central University, New Delhi, India, in 2007. She previously worked as a senior research fellow, research associate, and senior research associate (under the Scientists' Pool Scheme, Council of Scientific and Industrial Research (CSIR), India) at the Materials Research Laboratory, Department of Chemistry, JMI. Dr. Sharmin has more than fifty publications in peer-reviewed journals and books to her credit. She has presented more than thirty research papers at national and international conferences. Her research interests include the development of "green" materials with applications as antimicrobial and corrosion-resistant films, coatings, and packaging materials.

Contents

Preface

The bottom-up approach for the self-assembly of molecules, macromolecules, and particles into well-defined superstructures provides superior structural control of materials compared to top-down methods. Nature largely utilizes macromolecules to construct supramolecular materials, which ultimately contribute to the great array of forms and functions of life. Thus, the self-assembly of materials and the formation of superstructures have been of great interest in the fields of materials science, nanoscience, and nanoengineering.

This book describes the self-assembly of materials and supramolecular chemistry design principles for a broad spectrum of materials, including bio-inspired amphiphiles, metal oxides, metal nanoparticles, and organic–inorganic hybrid materials. It provides fundamental concepts of self-assembly design approaches and supramolecular chemistry principles for research ideas in nanotechnology applications. Written by leading scientists in their fields, this book includes six chapters organized into three sections: "Introduction to Self-Assembly and Supramolecular Chemistry", "Amphiphiles Self-Assembly", and "Nanoparticles Self-Assembly and Their Applications". We hope that this book will provide a foundation on supramolecular chemistry principles to students and active researchers who are interested in nanoscience and nanoengineering.

Hemali Rathnayake and Gayani Pathiraja
Nanoscience Department,
University of North Carolina at Greensboro,
Greensboro, USA

Eram Sharmin
Department of Pharmaceutical Chemistry,
College of Pharmacy,
Umm Al-Qura University,
Mecca, Saudi Arabia

Section 1

Introduction to Self-Assembly and Supramolecular Chemistry

Chapter 1

Introductory Chapter: Self-Assembly of Molecules into Supramolecular Structures

Hemali Rathnayake

1. Overview

The bottom-up approach for the self-assembly of molecules, macromolecules, and particles into well-defined superstructures provides superior structural control of materials compared to top-down methods. Nature largely utilizes macromolecules to construct supramolecular materials, ultimately contributing to a wide range of applications. Thus, the self-assembly of materials and the formation of superstructures have been of great interest in the fields of materials science, nanoscience, and nanoengineering.

This book provides the self-assembly of materials and supramolecular chemistry design principles for a broad spectrum of materials, including bio-inspired amphiphiles, metal oxides, metal nanoparticles, and organic-inorganic hybrid materials. It describes the fundamental concepts of self-assembly design approaches and supramolecular chemistry principles for research ideas in nanotechnology-enabled applications. The book focuses on three main themes, which include: the self-assembly and supramolecular chemistry of amphiphiles, the supramolecular structures and devices of inorganic materials, and the assembly-disassembly of organic-inorganic hybrid materials.

2. Self-assembly of bioinspired amphiphiles

Assembly of amphiphilic molecular and polymeric materials into precise functional structures is being explored actively for patterning and fabricating periodic array structures from mesoscale to nanoscale. Leading examples can be found in biosystems where assemblies of different amphiphilic components yield unusual microstructures that enable the performance of highly specific cellular functions [1–3]. For example, cell membrane's lipid bilayer, which is constructed from self-assembled phospholipids (PLs), is one of nature's well-organized nanoscale machines and is vital to cell's signal transduction. The self-assembly of biologically based lipids into unusual microstructures provides superior structural control due to the amphiphilic nature of lipids that are composed of a polar head group and a nonpolar hydrocarbon tail [4]. The dimensionality of the hydrated hydrophilic head group (0.7 to 1.0 nm) and the hydrophobic core of the bilayer (2.5 to 3.5 nm) results in varied tail lengths, which provide unprecedented control over the structural complexity and function at the nanoscale. For instance, most long-chain PLs self-assemble into spherical bilayer aggregates, known as liposomes [5]. However, by exploiting the subtle interplay between appropriately modified head groups and fatty acid chains, certain synthetic phospholipids self-assemble into novel microstructures [6]. One such example consists of synthetic PLs with photopolymerizable diacetylenic moieties in the acyl chains that self-assemble into hollow, cylindrical structures, known as tubules [7, 8], which have also been observed in other synthetic surfactants [9] and in bile [10].

Owing to these remarkable supramolecular assemblies, bio-conjugated nanomaterials, consisting of lipid bilayers and inorganic nanomaterials, such as metal nanoparticles [11–13] and nanocarbon [14, 15], have been demonstrated as new functional materials for electronic devices. In these devices, lipid bilayers are used as electrically insulating substrates that confine the nanomaterials. Particularly, the recent research advancements in this field set forth the supramolecular design principles, enabling access to many of the current examples of nanomaterials conjugated lipid bilayer architectures [11, 16–21]. These efforts toward biological mimicry of supramolecular assemblies suggest that metal-nanoparticle arrays can be self-assembled in the lipid bilayer vesicles, and one can separate an array of nanoparticles with a uniform gap distance through supramolecular principles. By controlling the separation distance between nanoparticles or arrays with nanoscale precision using selective molecular interactions of lipids, many variations in the size and shape of the assembled nanostructures are possible [22, 23]. Additionally, it has been demonstrated that lipids can alter the dielectric properties of metallic nanoparticles, providing a way to modulate the optical properties of an integrated architecture and serve as optical sensors [24]. It has also been proven that the interaction between the lipid and nanoparticle can be adjusted by altering the nature of the polar head groups [25]. Because the study of PLs assembly is used to model biological membranes [26], current studies of the hybrid structures of metallic nanoparticles and PLs have shown the potential for the development of new biosensing devices and drug delivery methods across cellular membranes.

Utilizing bioinspired molecular and polymeric systems, stimuli-responsive nanomaterials with a variety of functionalities that respond to light [27], pH [28], temperature [29], and chemical [30] stimuli have been also developed. Although a variety of synthetic strategies enables the development of responsive materials, their structures exhibit limitations in morphological deformation in response to external stimuli. Transition metal ions are a unique tool for engineering responsive character based on their various binding stoichiometries and geometries, allowing a single material to have a diverse set of responses to different metal ions.[7] Utilizing the dynamic nature of transition metal coordination bonds, structural control in small molecules [31–34] and proteins, [35–37] to develop responsive films, [38, 39] self-healing soft materials, [39–42] subcomponent self-assembly of polymeric materials, [43–46] and hierarchical assemblies of nanoparticles has been demonstrated [47, 48].

3. Supramolecular nanoassemblies of π-conjugated molecular systems

Conjugated organic molecules have been the subject of continuous interest in organic electronics. Synthetic efforts aimed at π-conjugated systems having well-defined architectures are indeed driven by the desire to impart specific optical and electrical properties to materials by controlling their molecular structure. Crafting the structure and function of organic materials using the strategies of self-assembly and supramolecular chemistry has progressed over the past two decades in materials chemistry. A variety of systems have been engineered where function is directly linked to non-covalent interactions, such as ionic, hydrophobic, van der Waals, hydrogen, and coordination bonds. They also influence solid-state self-organization to produce an organized structure at any scale, ranging from the nano and micrometer scales to microscopic dimensions.

Great efforts have been directed to the solution-processable self-assembly of π- conjugated small molecules, oligomers, or polymers into shape-defined nanostructures. Bridging the gap between natural and artificial systems, well-organized nanomaterials can be prepared using self-assembly approaches [49–54]. The morphologies of these self-assembled nanostructures cooperatively control by non-covalent interactions, such as H- bonding, dipole-dipole attraction, π-π stacking, van der Waals force, hydrophobic effect, electrostatic interaction, and metal-ligand coordination. Most predominantly, intermolecular π-π interactions and H-bonding have been the major driving forces, which often cooperatively drive other weak non-covalent interactions [50, 55]. Moreover, non-covalent interactions are highly dependent on the molecular structure where external environmental parameters, such as solvent, temperature, concentration, and fabrication process are responsible for the morphology of the supramolecular structure [50, 52, 56].

1D nanostructures of organic semiconductors with morphologies, including nanowires, nanobelts, nanorods, nanotubes, and helices, have merited intensive study over the past two decades. These 1D functional nanostructures hold great potential for enabling next-generation electronic and optoelectronic nanodevices [57, 58]. Consequently, the self-assembly of π-conjugated organic molecules into 1D nanostructures has been an active and rapidly developing field [52, 59–64].

In recent years, the application of Watson-Crick pairing of nucleic acids, that is, the pairing of nucleic acids, using specific hydrogen bonding to pattern self-assembling and supramolecular organic materials, has transformed how we design, engineer, and synthesize structures at all scales [65–68]. In the context of organic materials, the ability to generate ordered one-dimensional structures (1D) is fundamentally useful and functional. Thus, there remains considerable untapped potential in 1D organic materials. Successful fabrication of 1D nanostructures demands a tight correlation between the self-assembling kinetics and the molecular design and engineering. This usually requires a strong interplay between chemical synthesis, materials fabrication, and physical characterization that relate to a broad range of applications in electronic devices. Thus, understanding how to design molecules that can form 1D structures through non-covalent interactions is a key objective in organic material chemistry. So far, there is no universal principle that can predict the formation of any sort of 1D molecular self-assembly. Therefore, in order to design self-assembled 1D nanostructures, the supramolecular chemistry approaches should not only focus on the design principles of the molecular and supramolecular structure but also control the dynamic assembly-disassembly of the architecture. For example, they can be cylindrical assemblies, flat or twisted ribbons, tubes, and many other shapes but retain their morphologies in one dimension during the assembly and disassembly process.

Planar and rigid aromatic molecules are known to form one-dimensional (1D) and three-dimensional (3D) nanostructures through strong π-π interactions, evidencing that such interactions could be an effective approach for the formation of 1D nanoassemblies [69–80]. However, it still remains a difficult task to fabricate 1D nanostructures with well-defined morphology and molecular arrangement of these systems. Although impressive research have been performed on covalent and hydrogen-bonded dyads, triads, and higher order polycyclic aromatic systems in solution and solid phase [69–80]. research on further self-assembly of these complexes utilizing other strong non-covalent interactions, in particular, a combination of σ-holes and π–π interactions, is relatively scarce. This is quite unfortunate since self-assembly may have enormous potential for organic electronics, by providing a pathway to arrange electron donating (donors) and electron-accepting (acceptors) semiconducting fused-arene moieties into higher architectures through a wide variety of functional groups. In the solution phase, these systems can utilize for studying energy transfer phenomena as well as for understanding the feasibility of making dynamic self-assembled supramolecular structures by tailoring supramolecular interactions. For example, synergistic effect of non-covalent interactions, such as π-π stacking and hydrogen bonding on the chromophore stacking can reveal from their solution phase assembly behavior. These studies eventually can lead to the transfer of shape-persistent structures from solution to the thin film of the active layer of organic devices. Regardless, it is necessary to be programmed the molecular building blocks for the successful implementation of supramolecular electronics in practice, where the design enables the organization of functional chromophores into ordered, nano-sized aggregates.

Conflict of interest

The authors declare no conflict of interest.

Author details

Hemali Rathnayake
Nanoscience Department, University of North Carolina Greensboro, Greensboro, NC, USA

*Address all correspondence to: hprathna@uncg.edu

References

[1] Butler A, Theisen RM. Iron(III)–siderophore coordination chemistry: Reactivity of marine siderophores. Coordination Chemistry Reviews. 2010;**254**(3-4):288-296. Available from: https://pubmed.ncbi.nlm.nih.gov/21442004/

[2] Knight AS, Larsson J, Ren JM, Bou Zerdan R, Seguin S, Vrahas R, et al. Control of amphiphile self-assembly via bioinspired metal ion coordination. Journal of the American Chemical Society. 2018;**140**(4):1409-1414. Available from: https://europepmc.org/article/med/29338205

[3] Glotzer SC, Solomon MJ. Anisotropy of building blocks and their assembly into complex structures. Nature Materials. 2007;**6**(8):557-562. Available from: https://pubmed.ncbi.nlm.nih.gov/17667968/

[4] Katsaras J, Gutberlet T. Lipid Bilayers. Berlin, Heidelberg: Springer Berlin Heidelberg; 2001. DOI: 10.1007/978-3-662-04496-4

[5] Seddon JM, Templer RH. Polymorphism of lipid-water systems. Handbook of Biological Physics. 1995;**1995**:97-160. Available from: https://linkinghub.elsevier.com/retrieve/pii/S1383812106800205

[6] Lee YL, Lin JY, Chang CH. Thermodynamic characteristics and Langmuir–Blodgett deposition behavior of mixed DPPA/DPPC monolayers at air/liquid interfaces. Journal of Colloid and Interface Science. 2006;**296**(2):647-654. Available from: https://pubmed.ncbi.nlm.nih.gov/16225885/

[7] Bangham AD, Hill MW, Miller NGA. Preparation and use of liposomes as models of biological membranes. In: Methods in Membrane Biology. Boston, MA: Springer US; 1974. DOI: 10.1007/978-1-4615-7422-4_1

[8] Schnur JM, Peckerar MC, Stratton HM. Synthetic Microstructures in Biological Research. Boston, MA: Springer US; 1992. p. 249. Available from: https://agris.fao.org/agris-search/search.do?recordID=US201300721622

[9] Yager P, Schoen PE. Formation of tubules by a polymerizable surfactant. Molecular Crystals and Liquid Crystals. 1984;**106**(3-4):371-381. DOI: 10.1080/00268948408071454

[10] Schnur JM. Lipid tubules: A paradigm for molecularly engineered structures. Science. 1993;**262**(5140):1669-1676. DOI: 10.1126/science.262.5140.1669

[11] Nakashima N, Asakuma S, Kunitake T. Optical microscopic study of helical superstructures of chiral bilayer membranes. Journal of the American Chemical Society. 1985;**107**(2):509-510. DOI: 10.1021/ja00288a043

[12] Chung DS, Benedek GB, Konikoff FM, Donovan JM. Elastic free energy of anisotropic helical ribbons as metastable intermediates in the crystallization of cholesterol. Proceedings of the National Academy of Sciences. 1993;**90**(23):11341-11345. DOI: 10.1021/ja00288a043

[13] Song J, Cheng L, Liu A, Yin J, Kuang M, Duan H. Plasmonic vesicles of amphiphilic gold nanocrystals: Self-assembly and external-stimuli-triggered destruction. Journal of the American Chemical Society. 2011;**133**(28):10760-10763. DOI: 10.1021/ja204387w

[14] Song J, Zhou J, Duan H. Self-assembled plasmonic vesicles of SERS-encoded amphiphilic gold nanoparticles for cancer cell targeting and traceable intracellular drug delivery. Journal of the American Chemical Society. 2012;**134**(32):13458-13469. DOI: 10.1021/ja305154a

[15] Zeng Y, Zhang D, Wu M, Liu Y, Zhang X, Li L, et al. Lipid-AuNPs@PDA nanohybrid for MRI/CT imaging and photothermal therapy of hepatocellular carcinoma. ACS Applied Materials & Interfaces. 2014;**6**(16):14266-14277. Available from: https://pubmed.ncbi.nlm.nih.gov/25090604/

[16] Huang SCJ, Artyukhin AB, Misra N, Martinez JA, Stroeve PA, Grigoropoulos CP, et al. Carbon nanotube transistor controlled by a biological ion pump gate. Nano Letters. 2010;**10**(5):1812-1816. Available from: https://pubmed.ncbi.nlm.nih.gov/20426455/

[17] Frost R, Svedhem S, Langhammer C, Kasemo B. Graphene oxide and lipid membranes: Size-dependent interactions. Langmuir. 2016;**32**(11):2708-2717. DOI: 10.1021/acs.langmuir.5b03239

[18] Kim K, Geng J, Tunuguntla R, Comolli LR, Grigoropoulos CP, Ajo-Franklin CM, et al. Osmotically-driven transport in carbon nanotube porins. Nano Letters. 2014;**14**(12):7051-7056. Available from: https://pubmed.ncbi.nlm.nih.gov/25372973/

[19] Rasch MR, Rossinyol E, Hueso JL, Goodfellow BW, Arbiol J, Korgel BA. Hydrophobic gold nanoparticle self-assembly with phosphatidylcholine lipid: Membrane-loaded and Janus vesicles. Nano Letters. 2010;**10**(9):3733-3739. DOI: 10.1021/nl102387n

[20] Rasch MR, Yu Y, Bosoy C, Goodfellow BW, Korgel BA. Chloroform-enhanced incorporation of hydrophobic gold nanocrystals into dioleoylphosphatidylcholine (DOPC) vesicle membranes. Langmuir. 2012;**28**(36):12971-12981. Available from: https://www.ncbi.nlm.nih.gov/pmc/articles/pmid/22897240/?tool=EBI

[21] Šegota S, Vojta D, Kendziora D, Ahmed I, Fruk L, Baranović G. Ligand-dependent nanoparticle clustering within lipid membranes induced by surrounding medium. The Journal of Physical Chemistry. B. 2015;**119**(16):5208-5219. DOI: 10.1021/acs.jpcb.5b00898

[22] Park SH, Oh SG, Mun JY, Han SS. Loading of gold nanoparticles inside the DPPC bilayers of liposome and their effects on membrane fluidities. Colloids and Surfaces. B, Biointerfaces. 2006;**48**(2):112-118. Available from: https://pubmed.ncbi.nlm.nih.gov/16520025/

[23] Sakaguchi N, Kimura Y, Hirano-Iwata A, Ogino T. Fabrication of Au-nanoparticle-embedded lipid bilayer membranes supported on solid substrates. The Journal of Physical Chemistry. B. 2017;**121**(17):4474-4481. Available from: https://pubmed.ncbi.nlm.nih.gov/28414450/

[24] Howland MC, Szmodis AW, Sanii B, Parikh AN. Characterization of physical properties of supported phospholipid membranes using imaging ellipsometry at optical wavelengths. Biophysical Journal. 2007;**92**(4):1306-1317. Available from: https://linkinghub.elsevier.com/retrieve/pii/S0006349507709414

[25] Nieh MP, Raghunathan VA, Glinka CJ, Harroun TA, Pabst G, Katsaras J. Magnetically alignable phase of phospholipid "Bicelle" mixtures is a chiral nematic made up of wormlike micelles. Langmuir.

2004;**20**(19):7893-7897. Available from: https://pubmed.ncbi.nlm.nih.gov/15350048/

[26] Ma G, Allen HC. Condensing effect of palmitic acid on DPPC in mixed langmuir monolayers. Langmuir. 2007;**23**(2):589-597. Available from: https://pubmed.ncbi.nlm.nih.gov/17209610/

[27] Coleman AC, Beierle JM, Stuart MCA, Maciá B, Caroli G, Mika JT, et al. Light–induced disassembly of self-assembled vesicle-capped nanotubes observed in real time. Nature Nanotechnology. 2011;**6**(9):547-552. Available from: https://pubmed.ncbi.nlm.nih.gov/21841795/

[28] Jang SG, Audus DJ, Klinger D, Krogstad DV, Kim BJ, Cameron A, et al. Striped, ellipsoidal particles by controlled assembly of diblock copolymers. Journal of the American Chemical Society. 2013;**135**(17):6649-6657. DOI: 10.1021/ja4019447

[29] Moughton AO, O'Reilly RK. Thermally induced micelle to vesicle morphology transition for a charged chain end diblock copolymer. Chemical Communications. 2010;**46**(7):1091-1093. Available from: https://pubs.rsc.org/en/content/articlehtml/2010/cc/b922289h

[30] Yan Q, Zhao Y. CO_2-stimulated diversiform deformations of polymer assemblies. Journal of the American Chemical Society. 2013;**135**(44):16300-16303. DOI: 10.1021/ja408655n

[31] Philip A, Gale JWS. Supramolecular Chemistry: From Molecules to Nanomaterials. 1st ed. Chichester, West Sussex: Wiley; 2012. p. 4014. Available from: https://www.worldcat.org/title/753634033

[32] Geng Y, Wang XJ, Chen B, Xue H, Zhao YP, Lee S, et al. Semiconducting neutral microstructures fabricated by coordinative self-assembly of intramolecular charge-transfer tetrathiafulvalene derivatives. Chemistry - A European Journal. 2009;**15**(20):5124-5129. Available from: https://pubmed.ncbi.nlm.nih.gov/19322846/

[33] Li L, Ke CF, Zhang HY, Liu Y. Coordination-induced switchable nanoparticle formation from naphthyl-bridged bis(β-cyclodextrin). The Journal of Organic Chemistry. 2010;**75**(19):6673-6676. Available from: https://pubmed.ncbi.nlm.nih.gov/20806949/

[34] Xiao ZY, Zhao X, Jiang XK, Li ZT. Tunable coordinative assembly of a disc-like molecule and metal ions: From mirospheres to microtubes and microrods. Chemistry of Materials. 2011;**23**(6):1505-1511. DOI: 10.1021/cm103182e

[35] van Eldijk MB, Schoonen L, Cornelissen JJLM, Nolte RJM, van Hest JCM. Metal ion-induced self-assembly of a multi-responsive block copolypeptide into well-defined nanocapsules. Small. 2016;**12**(18):2476-2483. Available from: https://pubmed.ncbi.nlm.nih.gov/27151830/

[36] Salgado EN, Radford RJ, Tezcan FA. Metal-directed protein self-assembly. Accounts of Chemical Research. 2010;**43**(5):661-672. DOI: 10.1021/ar900273t

[37] Nepal M, Sheedlo MJ, Das C, Chmielewski J. Accessing three-dimensional crystals with incorporated guests through metal-directed coiled-coil peptide assembly. Journal of the American Chemical Society. 2016;**138**(34):11051-11057. DOI: 10.1021/jacs.6b06708

[38] Michal BT, McKenzie BM, Felder SE, Rowan SJ. Metallo-, thermo-,

and photoresponsive shape memory and actuating liquid crystalline elastomers. Macromolecules. 2015;**48**(10):3239-3246. DOI: 10.1021/acs.macromol.5b00646

[39] Mozhdehi D, Ayala S, Cromwell OR, Guan Z. Self-healing multiphase polymers via dynamic metal–ligand interactions. Journal of the American Chemical Society. 2014;**136**(46):16128-16131. DOI: 10.1021/ja5097094

[40] Grindy SC, Learsch R, Mozhdehi D, Cheng J, Barrett DG, Guan Z, et al. Control of hierarchical polymer mechanics with bioinspired metal-coordination dynamics. Nature Materials. 2015;**14**(12):1210-1216. Available from: https://pubmed.ncbi.nlm.nih.gov/26322715/

[41] Whittell GR, Hager MD, Schubert US, Manners I. Functional soft materials from metallopolymers and metallosupramolecular polymers. Nature Materials. 2011;**10**(3):176-188. Available from: https://pubmed.ncbi.nlm.nih.gov/21336298/

[42] Ott C, Hoogenboom R, Hoeppener S, Wouters D, Gohy JF, Schubert US. Tuning the morphologies of amphiphilic metallo-supramolecular triblock terpolymers: From spherical micelles to switchable vesicles. Soft Matter. 2009;**5**(1):84-91. Available from: https://pubs.rsc.org/en/content/articlehtml/2009/sm/b813161a

[43] Gao H, Liu G, Chen X, Hao Z, Tong J, Lu L, et al. Media-modulated interchain or intrachain coordination of amphiphilic block copolymer micelles. Macromolecules. 2010;**43**(14):6156-6165. DOI: 10.1021/ma100761a

[44] Chen X, Xu N, Li N, Lu L, Cai Y, Zhao Y, et al. Programmable selectivity of metal–imine bond coordination in subcomponent self-assembly of a primary amine based block copolymer. Soft Matter. 2013;**9**(6):1885-1894. Available from: https://pubs.rsc.org/en/content/articlehtml/2013/sm/c2sm26904j

[45] Wu X, Xu N, Zhu Z, Cai Y, Zhao Y, Wang D. Subcomponent self-assembly of polymer chains based on dynamic and geometrical coordination diversity of the first row transition metal ions. Polymer Chemistry. 2014;**5**(4):1202-1209. Available from: https://pubs.rsc.org/en/content/articlehtml/2014/py/c3py01111a

[46] Xu N, Han J, Zhu Z, Song B, Lu X, Cai Y. Directional supracolloidal self-assembly via dynamic covalent bonds and metal coordination. Soft Matter. 2015;**11**(27):5546-5553. Available from: https://pubs.rsc.org/en/content/articlehtml/2015/sm/c5sm00546a

[47] Wang Y, Hollingsworth AD, Yang SK, Patel S, Pine DJ, Weck M. Patchy particle self-assembly via metal coordination. Journal of the American Chemical Society. 2013;**135**(38):14064-14067. Available from: https://pubmed.ncbi.nlm.nih.gov/24044358/

[48] Lunn DJ, Gould OEC, Whittell GR, Armstrong DP, Mineart KP, Winnik MA, et al. Microfibres and macroscopic films from the coordination-driven hierarchical self-assembly of cylindrical micelles. Nature Communications. 2016;**7**(1):12371. Available from: https://pubmed.ncbi.nlm.nih.gov/27538877/

[49] Whitesides GM, Simanek EE, Mathias JP, Seto CT, Chin D, Mammen M, et al. Noncovalent synthesis: Using physical-organic chemistry to make aggregates. Accounts of Chemical Research. 1995;**28**(1):37-44. DOI: 10.1021/ar00049a006

[50] Hoeben FJM, Jonkheijm P, Meijer EW, Schenning APHJ. About

supramolecular assemblies of π-conjugated systems. Chemical Reviews. 2005;**105**(4):1491-1546. DOI: 10.1021/cr030070z

[51] Zhang Z, Ma R, Shi L. Cooperative macromolecular self-assembly toward polymeric assemblies with multiple and bioactive functions. Accounts of Chemical Research. 2014;**47**(4):1426-1437. DOI: 10.1021/ar5000264

[52] Schenning APHJ, Meijer EW. Supramolecular electronics; nanowires from self-assembled π-conjugated systems. Chemical Communications. 2005;**26**:3245. Available from: http://xlink.rsc.org/?DOI=b501804h

[53] Wu J, Pisula W, Müllen K. Graphenes as potential material for electronics. Chemical Reviews. 2007;**107**(3):718-747. DOI: 10.1021/cr068010r

[54] Würthner F. Supramolecular polymerization: Living it up. Nature Chemistry. 2014;**6**(3):171-173. Available from: https://pubmed.ncbi.nlm.nih.gov/24557127/

[55] González-Rodríguez D, Schenning APHJ. Hydrogen-bonded supramolecular π-functional materials. Chemistry of Materials. 2011;**23**(3):310-325. DOI: 10.1021/cm101817h

[56] Ariga K, Hill JP, Lee MV, Vinu A, Charvet R, Acharya S. Challenges and breakthroughs in recent research on self-assembly. Science and Technology of Advanced Materials. 2008;**9**(1):014109. DOI: 10.1088/1468-6996/9/1/014109

[57] Nguyen TQ, Martel R, Bushey M, Avouris P, Carlsen A, Nuckolls C, et al. Self-assembly of 1-D organic semiconductor nanostructures. Physiological Chemistry and Physics. 2007;**9**(13):1515. Available from: https://pubs.rsc.org/en/content/articlehtml/2007/cp/b609956d

[58] Kim FS, Ren G, Jenekhe SA. One-dimensional nanostructures of π-conjugated molecular systems: Assembly, properties, and applications from photovoltaics, sensors, and nanophotonics to nanoelectronics. Chemistry of Materials. 2011;**23**(3):682-732. DOI: 10.1021/cm102772x

[59] Meijer EW, Schenning APHJ. Material marriage in electronics. Nature. 2002;**419**(6905):353-354. Available from: https://pubmed.ncbi.nlm.nih.gov/12353020/

[60] Wang X, Li Y. Solution-based synthetic strategies for 1-D nanostructures. Inorganic Chemistry. 2006;**45**(19):7522-7534. DOI: 10.1021/ic051885o

[61] Zang L, Che Y, Moore JS. One-dimensional self-assembly of planar π-conjugated molecules: Adaptable building blocks for organic nanodevices. Accounts of Chemical Research. 2008;**41**(12):1596-1608. DOI: 10.1021/ar800030w

[62] Zhao YS, Fu H, Peng A, Ma Y, Liao Q, Yao J. Construction and optoelectronic properties of organic one-dimensional nanostructures. Accounts of Chemical Research. 2010;**43**(3):409-418. DOI: 10.1021/ar900219n

[63] Su B, Wu Y, Jiang L. The art of aligning one-dimensional (1D) nanostructures. Chemical Society Reviews. 2012;**41**(23):7832. Available from: https://pubs.rsc.org/en/content/articlehtml/2012/cs/c2cs35187k

[64] Dasgupta NP, Sun J, Liu C, Brittman S, Andrews SC, Lim J, et al. 25th Anniversary Article: Semiconductor nanowires - Synthesis, characterization,

and applications. Advanced Materials. 2014;**26**(14):2137-2184. DOI: 10.1002/adma.201305929

[65] Zheng J, Constantinou PE, Micheel C, Alivisatos AP, Kiehl RA, Seeman NC. Two-dimensional nanoparticle arrays show the organizational power of robust DNA motifs. Nano Letters. 2006;**6**(7):1502-1504. Available from: https://pubmed.ncbi.nlm.nih.gov/16834438/

[66] Park SY, Lytton-Jean AKR, Lee B, Weigand S, Schatz GC, Mirkin CA. DNA-programmable nanoparticle crystallization. Nature. 2008;**451**(7178):553-556. Available from: https://pubmed.ncbi.nlm.nih.gov/18235497/

[67] Aldaye FA, Palmer AL, Sleiman HF. Assembling materials with DNA as the guide. Science. 2008;**321**(5897):1795-1799. Available from: https://pubmed.ncbi.nlm.nih.gov/18818351/

[68] Sayar M, Stupp SI. Self-organization of rod-coil molecules into nanoaggregates: A coarse grained model. Macromolecules. 2001;**34**:7135-7139. Available from: https://pubs.acs.org/sharingguidelines

[69] Duzhko V, Singer KD. Self-assembled fibers of a discotic phthalocyanine derivative: Internal structure, tailoring of geometry, and alignment by a direct current electric field. Journal of Physical Chemistry C. 2007;**111**(1):27-31. DOI: 10.1021/jp066936x

[70] Hill JP, Jin W, Kosaka A, Fukushima T, Ichihara H, Shimomura T, et al. Self-assembled hexa- peri -hexabenzocoronene graphitic nanotube. Science. 2004;**304**(5676):1481-1483. DOI: 10.1126/science.1097789

[71] Briseno AL, Mannsfeld SCB, Reese C, Hancock JM, Xiong Y, Jenekhe SA, et al. Perylenediimide nanowires and their use in fabricating field-effect transistors and complementary inverters. Nano Letters. 2007;**7**(9):2847-2853. DOI: 10.1021/nl071495u

[72] Praefcke K, Singer D, Kohne B, Ebert M, Liebmann A, Wendorff JH. Charge transfer induced nematic columnar phase in low molecular weight disc-like systems. Liquid Crystals. 1991;**10**(2):147-159. DOI: 10.1080/02678299108036421

[73] Tang Q, Li H, He M, Hu W, Liu C, Chen K, et al. Low threshold voltage transistors based on individual single-crystalline submicrometer-sized ribbons of copper phthalocyanine. Advanced Materials. 2006;**18**(1):65-68. DOI: 10.1002/adma.200501654

[74] Bendikov M, Wudl F, Perepichka DF. Tetrathiafulvalenes, oligoacenenes, and their buckminsterfullerene derivatives: The brick and mortar of organic electronics. Chemical Reviews. 2004;**104**(11):4891-4946. Available from: https://pubmed.ncbi.nlm.nih.gov/15535637/

[75] Balakrishnan K, Datar A, Oitker R, Chen H, Zuo J, Zang L. Nanobelt self-assembly from an organic n-type semiconductor: Propoxyethyl-PTCDI. Journal of the American Chemical Society. 2005;**127**(30):10496-10497. DOI: 10.1021/ja052940

[76] Balakrishnan K, Datar A, Naddo T, Huang J, Oitker R, Yen M, et al. Effect of side-chain substituents on self-assembly of perylene diimide molecules: Morphology control. Journal of the American Chemical Society. 2006;**128**(22):7390-7398. Available from: https://pubmed.ncbi.nlm.nih.gov/16734495/

[77] Sinks LE, Rybtchinski B, Iimura M, Jones BA, Goshe AJ, Zuo X, et al. Self-assembly of photofunctional cylindrical

nanostructures based on perylene-3,4:9,10-bis(dicarboximide). Chemistry of Materials. 2005;**17**(25):6295-6303. DOI: 10.1021/cm051461s

[78] Gesquière A, Jonkheijm P, Hoeben FJM, Schenning APHJ, De Feyter S, De Schryver FC, et al. 2D-structures of quadruple hydrogen bonded oligo(p-phenylenevinylene)s on graphite: Self-assembly behavior and expression of chirality. Nano Letters. 2004;**4**(7):1175-1179. DOI: 10.1021/nl049842c

[79] Schenning APHJ et al. Photoinduced electron transfer in hydrogen-bonded oligo(p -phenylene vinylene)—perylene bisimide chiral assemblies. Journal of the American Chemical Society. 2002;**124**(35):10252-10253. DOI: 10.1021/ja020378s

[80] Gehan TS, Bag M, Renna LA, Shen X, Algaier DD, Lahti PM, et al. Multiscale active layer morphologies for organic photovoltaics through self-assembly of nanospheres. Nano Letters. 2014;**14**(9):5238-5243. DOI: 10.1021/nl502209s

Section 2

Amphiphiles Self-Assembly

Chapter 2

Porphyrin Self-Assembled Nanostructures and Applications

Shiqi Liu, Zengyin Li, Huiwei Tong, Yong Zhong and Feng Bai

Abstract

Porphyrins are a class of macromolecular heterocyclic compounds formed by the inter-carbon atoms of four pyrrole-like subunits through the submethyl bridge (=CH−). Porphyrin rings have 26 electrons in highly conjugated system and are easily modified peripheral structures, often serve as ideal building blocks to construct self-assembled nanostructures with excellent physical and chemical properties. Porphyrin nanostructures have excellent visible light absorption properties, which will significantly improve the efficiency of electron–hole separation, and are also commonly used in photocatalysis fields. Porphyrin photosensitizers have superior strong phototoxicity and little side effects, and are widely used in tumor photothermal/photodynamic treatment. This chapter summarizes the self-assembly methods of porphyrins, the applications progress of porphyrin self-assembled nanomaterials in photocatalysis and tumor therapy, and discusses the development trend in future of porphyrin nanomaterials.

Keywords: porphyrin, photosensitizer, self-assembly, photocatalytic, therapy

1. Introduction

Porphyrins are widely found in animals' blood and plants in nature. They are macromolecular heterocyclic compounds, have highly conjugated system, and are formed by the cross-carbon atoms of four pyrrole subunits through a submethyl bridge (=CH−) [1]. Because the rigid conjugated electron ring and adjustable geometry endow porphyrin molecule the self-assembly characteristics [2], conjugate molecular framework of delocalized aromatic electron ensure them excellent physical, chemical, photochemical, and biological characteristics, which make them applicate in supramolecular electronics, light collection, energy conversion system, chemical/biosensors, and biomedicine fields [3, 4].

Self-assembly process is that smaller structural units spontaneously assemble into nanostructures with confined morphology driven by various non-covalent intermolecular interactions, such as hydrogen bonds, π-π packing, hydrophobic interactions, electrostatic interactions, and van der Waals forces. Porphyrin self-assembled nanostructures will inherit the characteristics of porphyrin monomer, and possess electronic buffer, photoelectric conversion, photosensitivity, and high chemical stability, so they have been a hot object of research in photocatalysis and nanomedicine fields [5].

Molecular self-assembly is an effective method for preparing nanostructured materials [6]. The unique properties of porphyrin self-assembled nanostructures depend not only on the size, morphology, and composition of molecular units, but also on the large-scale spatial arrangement order of the assembly. Therefore, the construction of porphyrin self-assembled nanostructures is an important research direction in the current field of materials science. This chapter focused on the preparation methods of porphyrin self-assembly, characterization methods, and application fields of porphyrin self-assembled nanomaterials, and explore the future development trend of such materials.

2. Methods of porphyrin self-assembly

The preparation methods of porphyrin self-assembled nanostructures mainly include acid–base neutralization/micelle confined self-assembled method, surfactant-assisted mixing solvent assembly method, microemulsion-assisted/mesophase transfer self-assembly method, peptide or nanocrystal, and porphyrin coassembly method. Of course, there are other strategies, such as ion self-assembly [7] and droplet volatilization-induced self-assembly, which will not be described in this chapter.

2.1 Acid: base neutralization/micelle confined self-assembled method

Most porphyrin monomer is insoluble in water, which affects their wide application in catalysis, biotherapy, etc. In order to solve this problem, Bai developed an acid–base neutralization micelle confinement self-assembly method. They dissolved the pyridyl porphyrin in an acidic solution, or dissolved the phenol, phenylcarboxyl porphyrin, etc. in an alkaline solution, and then poured them into the alkaline or acidic solution of the emulsifier, and an acid–base neutralization reaction occurs (**Figure 1A**), and the porphyrin molecules are reprecipitated and self-assembled to form a regular and orderly stacked self-assembled nanostructure (**Figure 1B**) [8]. Therefore, the pH of the solution and the ratio of porphyrin emulsifier, and assemblies of different sizes can be obtained [9].

Typically, 5,10,15,20-tetrakis(4-(hydroxyl)phenyl) porphyrin (THPP) was dissolved in NaOH aqueous solution forming a water-soluble TPP^{4-} anion, then was quickly poured into an acidic emulsifier solution ($c_{surfactant}$ > critical micelle concentration (CMC)), instantaneously occur acid–base neutralization, then TPP^{4-} anion protonation and forms neutral THPP reprecipitation, and entered into the emulsifier micellar, after stirring agitation and growth, THPP molecule self-assembled into nanowires driven by the non-covalent interactions, such as hydrogen bond, π-π interaction, hydrophobic interaction, and van der Waals force. Scanning electron microscopy (SEM) (**Figure 2A**) shows the average length and diameter of THPP nanowires and are about 4.5 μm and 110 nm, respectively. Besides, according to the transmission electron microscopy (TEM) images (**Figure 2B**) and high-resolution TEM (HR-TEM) image (**Figure 2C**), the nanowires are uniform and monodispersed without defects with well-resolved lattice fringes with interplanar distance of 1.39 nm, indicating the possible formation of multiple hydrogen bonds, such as (pyrrol) N + H···Cl and OH···H, among THPP molecules within the nanowires (**Figure 2D**), which are balanced favoring the formation of J-aggregates [8].

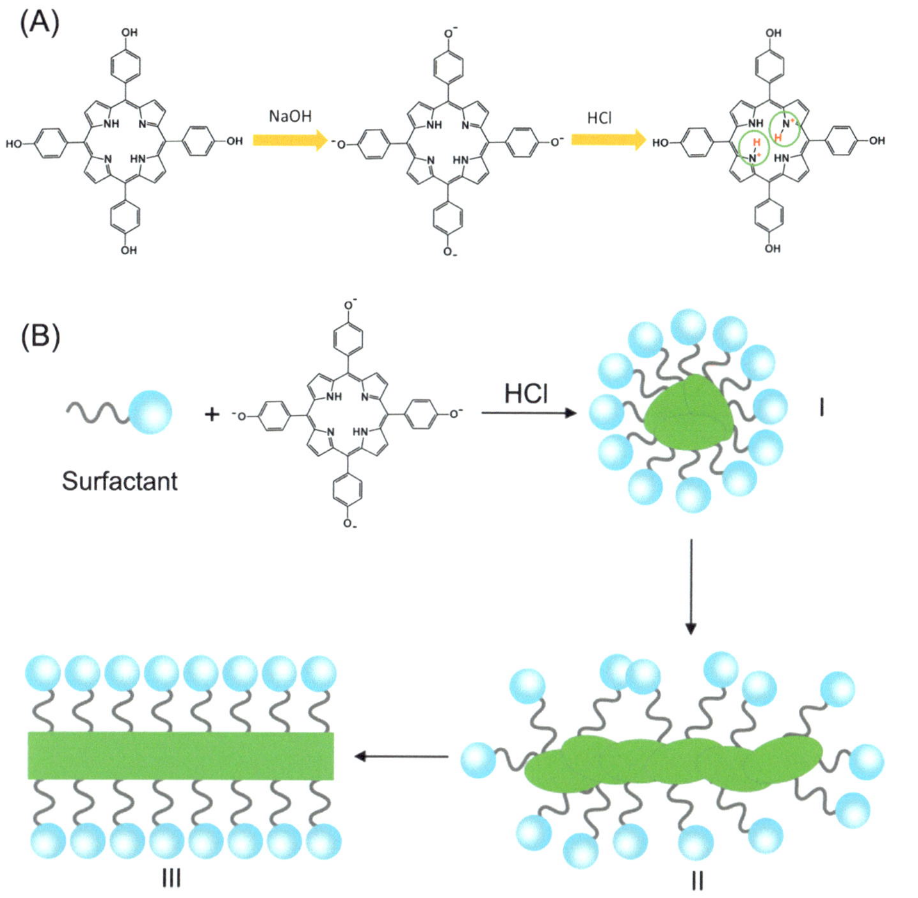

Figure 1.
(A) Scheme of THPP dissolved in NaOH aqueous solution and protonation process. (B) Schematic illustration of acid–base neutralization micelle confined self-assembled method. Reproduced with permission from reference [8]. Copyright 2018 American Chemical Society.

The pyridine and pyrrole groups are inclined to protonated and dissolve in large amounts in an acidic aqueous solution, lead to pyridine matrix and form MTPyP-H_4^{4+} soluble tetrapyridinium solution [9]. Then, the acidification porphyrin solution was injected into a basic surfactant aqueous solution ($c_{surfactant}$ > CMC) under vigorous stirring at room temperature (25°C), immediately triggers an acid–base neutralization reaction. Then, the deprotonation of MTPyP-H_4^{4+} produced neutral insoluble MTPyP porphyrins and encapsulated into the hydrophobic cores of surfactant micelles. Driven by intermolecular non-covalent interactions, such as π-π stacking, hydrogen bonding, Zn-N axial coordination, hydrophobic-hydrophobic interactions, and so on. MTPyP molecules started the self-assembly process with the assistance of surfactant template and further initiated nucleation and growth to form uniform porphyrin nanocrystals.

Based on this, the self-assembly process of Co-tetra(4-pyridyl) porphyrin (CoTPyP) underwent very slow, and hardly get self-assembled product less than 10 days at 25°C. After continuous stirring and aging for 100 days [10], a small amount of CoTPyP regular polyhedral nanoparticles with an average size 300 nm were obtained (**Figure 2E**). Until continuous stirring for 150 days, the CoTPyP monomers

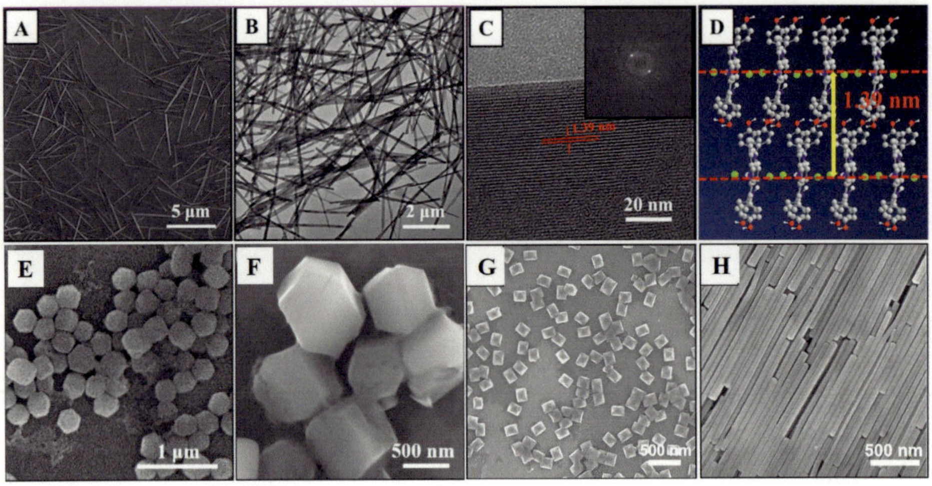

Figure 2.
Structure characterizations of the self-assembled porphyrin nanostructures by acid–base neutralization micelle confined self-assembly method. (A) SEM image of the THPP nanowires. (B) Corresponding TEM and (C) HR-TEM image of (A), the inset is FFT. (D) Crystal structure simulated of THPP nanowires (chloride, green; nitrogen, blue; oxygen, red; carbon, gray). Reproduced with permission from reference [8]. Copyright 2018 American Chemical Society. SEM images of CoTPyP nanocrystals were collected at (E) 100 d and (F) 150d. Reproduced with permission from reference [10]. Copyright 2020 The Materials Research Society. (G) SEM of the self-assembled H_2TPyP nanooctahedra. (H) SEM of the self-assembled Zn-metalated H_2TPyP nanowires. Reproduced with permission from reference [11]. Copyright 2016 American Chemical Society.

completed the self-assembly process and formed uniform and regular hexagonal prism nanocrystals. The size of CoTPyP prism is 700 nm × 1000 nm × 750 nm (**Figure 2F**), and the interplanar distance of lattice fringes is 1.53 nm. Contrarily, the meso-tetra(4-pyridyl) porphine (H_2TPyP) can self-assemble into well-defined 3D octahedral morphology (**Figure 2G**) with no apparent defects in 12 h. Moreover, the self-assembled structures obtained by different central metalloporphyrins are also different. When Zn^{2+} (through $Zn(NO_3)_2$) was added in the H_2TPyP self-assembly solution, Zn-metalation into the core of H_2TPyP forming ZnTPyP [11], the morphology turned in nanowires (**Figure 2H**). The porphyrin self-assembled structure synthesized based on this method can be easily controlled by the solution pH, the concentration of the emulsifier, the concentration of the porphyrin, etc., and can be prepared in large amounts.

2.2 Surfactant-assisted mixing solvent assembly method

The good solvent solution of porphyrin is poured into the poor solvent of porphyrin, and the porphyrin reprecipitates and assembles to form nanostructures with the surfactant-assisted due to the change of solubility. Therefore, the size and morphology of self-assembled nanostructures can be controlled by the solvent polarity and the mole ratio of surfactant and porphyrin.

Typically, Pd (II) tetra (4-carboxylphenyl) porphyrin (PdTCPP) powder is insoluble in water but has good solubility in N, N-dimethylformamide (DMF) solution. Therefore, PdTCPP powder dissolved in DMF to obtain a PdTCPP/DMF homogeneous solution, and then quickly added into to the 1-decanesulfonic acid sodium salt (DASS) aqueous solution at one time under stirring at room temperature. Instantly, PdTCPP molecules are precipitated from the DMF system and started to self-assemble driven by hydrophobic-hydrophobic interaction between porphyrin molecules and the end of the surfactant and π-π stacking and hydrogen bonding [12]. The SEM and TEM images of the PdTCPP assemblies showed that these assemblies are two-dimensional leaf-like nanostructures with highly regular morphology and uniform size (**Figure 3A**), and the surface of the PdTCPP nanoleaf is smooth without obvious defects. The fine structure of TEM showed that the edge of PdTCPP nanoleaf showed a jagged irregular shape, and the structure was superimposed layer by layer (**Figure 3B**). The AFM results further demonstrated that the surface of the PdTCPP nanoleaf was smooth and the thickness was about 200 nm (**Figure 3C** and **D**). The long and short axes of the PdTCPP nanoleaf are 5.43 μm and 1.64 μm, respectively.

In order to acquire the internal fine structure, the PdTCPP nanoleaves were embedded in epoxy resin and made ultrathin sections. The cross-section of the PdTCPP nanoleaves was rectangular and uniformly distributed in the slice, and the thickness of the vertical interface is about 200 nm (**Figure 3E**). HR-TEM of cross-sectional showed that there was 1.5 nm distinct and well-resolved lattice fringe on the cross-section of the PdTCPP nanoleaf (**Figure 3F**). Moreover, the obvious diffraction spots of the layered structure are obtained by fast Fourier transform (FFT), indicating that the PdTCPP nanoleaf has a good crystal structure (**Figure 3F**). The XRD test showed that the films and powders of PdTCPP nanoleaf had strong diffraction peaks at 5.95, 11.67, 17.39, 23.16, and 28.97° (**Figure 3G**), and the positions of the diffraction peaks were approximately 1:2:3:4:5, which also proved that the PdTCPP nanoleaf is a layer-by-layer structure. While the commercial PdTCPP powder has no diffraction peak as a control. The density functional theory calculation of the crystal structure shows that the PdTCPP nanoleaf is mainly formed through an orderly arrangement of porphyrin intermolecular by the hydrogen bonding and π-π stacking interaction. (**Figure 3H**).

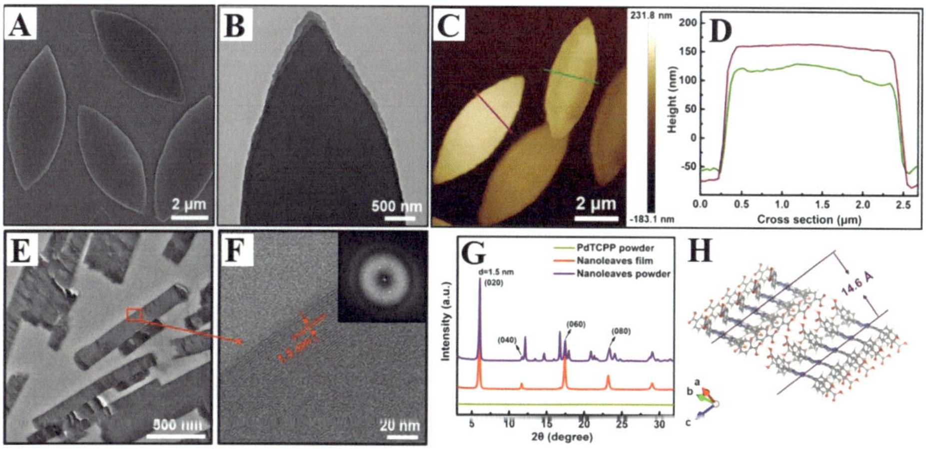

Figure 3.
The self-assembled PdTCPP nanoleaves were prepared by surfactant-assisted mixing solvent assembly method. (A) SEM image, (B) TEM image, and (C) AFM image of the nanoleaves. (D) the thickness of the vertical interface in panel (C). (E) TEM image of the cross-section, (F) HRTEM image of the corresponding (E); the inset is the corresponding FFT, (G) XRD patterns, (H) simulated crystal structure of the nanoleaves. Reproduced with permission from reference [12]. Copyright 2022 American Chemical Society.

2.3 Microemulsion-assisted/mesophase transfer self-assembly method

Microemulsion refers to a thermodynamically stable, isotropic, transparent, or translucent dispersion system formed by two immiscible liquids in the presence of surfactants, and their particle sizes distribution is within 10–100 nm. The microemulsion system is composed of oil phase, water phase, and surfactant [13]. Nanoparticles prepared by microemulsion technology have a controllable shape and size, relatively narrow particle size distribution, and hard to agglomerate [14, 15]. In the process of preparing microemulsion, the oil phase and the water phase are emulsified under ultrasonic probe, and microemulsion droplets with narrow size distribution are obtained. These droplets can exist stably in the aqueous phase in the presence of surfactants or stabilizers and can act as nanoreactors for the reaction. This "nanoreactor" can change the particle size according to the number of solubilized substances in the droplet. The more solubilized substances, the larger the radius of the reaction core, and the larger the radius of the generated nanoparticles, which can control the shape and size in the preparation of the nanoparticles process.

Liu et al. reported a microemulsion droplet nanoreactor using cetyltrimethylammonium bromide (CTAB) as a surfactant. Typically, the $CHCl_3$ solution of In (III) mesotetraphenyl porphine chloride (InTPP) was injected into CTAB aqueous solution. Then, the reaction solution was placed under an ultrasonic probe to emulsify

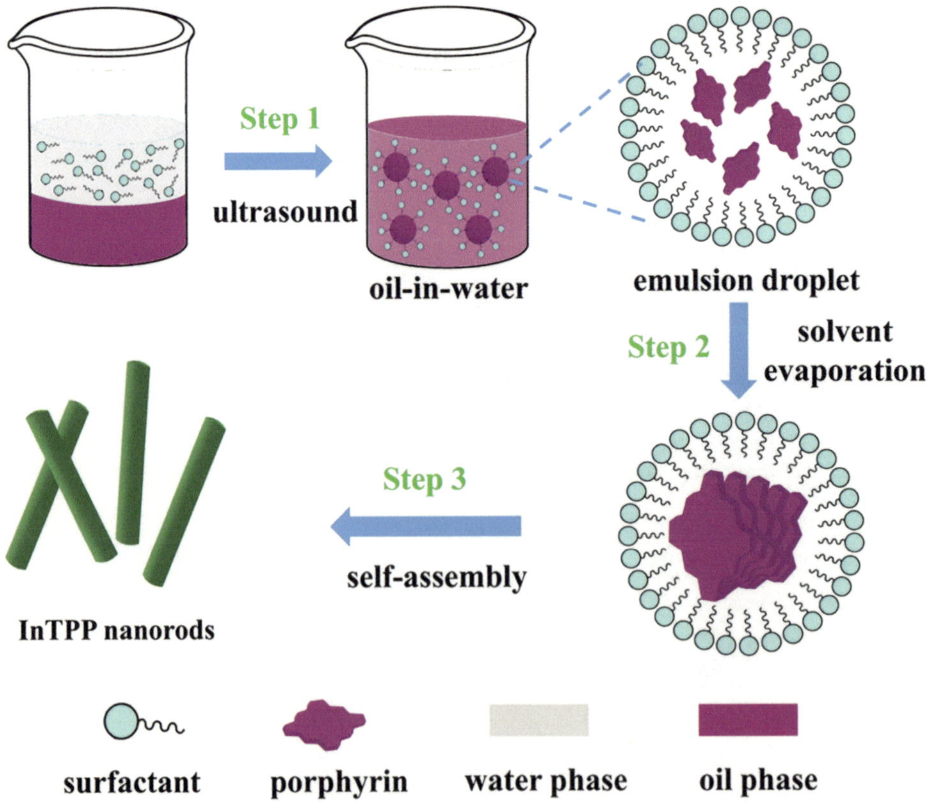

Figure 4.
Schematic illustration of the microemulsion-assisted/mesophase transfer self-assembly process. Reproduced with permission from reference [16]. Copyright 2019 American Chemical Society.

and sonicated at 80 W for 1 min. The microemulsion then transfers in a 60°C water bath to volatilize chloroform solvent for 30 min, the InTPP molecules self-assembled into regular nanostructures driven by non-covalent interactions (**Figure 4**). After that, the porphyrin solution was centrifuged to separate the precipitate, then washed with water to remove excess surfactant. By changing the mole ratio of porphyrin and surfactant, the size and morphology of the resulting assemblies can be regulated. Starting from here, InTPP self-assembled nanorods were successfully synthesized depending on the concentrations of CTAB [16]. Representative TEM images show that the resulted InTPP nanocrystals are uniform and monodisperse nanowires or nanorods (**Figure 5A–D**). Moreover, the well-defined nanowires or nanorods with controlled aspect ratios of 81.3, 21.0, 4.6, and 1.6 can be adjusted by the CTAB concentration. These results further confirmed that the surfactant microemulsion droplets are good nanoreactors to control the size of nanoparticles.

Similarly, the microemulsion droplet approach is also applicable to the self-assembly of phthalocyanine nanocrystals. As shown in **Figure 5E**, when the emulsifier is sodium dodecyl sulfate (SDS), the morphology of the manganese phthalocyanine (MnPc) assemblies is mainly spherical nanoparticles with a diameter of ~62 nm, and the size changes with the SDS concentrations. When sodium dodecyl benzene sulfonate (SDBS) was used, the morphologies of the MnPc assemblies were all linear (**Figure 5F**) with a 1 μm in length. With the decrease of the emulsifier concentration, the MnPc assemblies gradually grew from linear (0.02 M) to sheet-like assemblies (0.01 M), further reducing the concentration of SDBS (below 0.005 M), the length of the assembly gradually decreased [17].

Comparably, the Au(III) tetra-(4-pyridyl) porphine (AuTPyP) can self-assemble into nanospheres with an average diameter of ~65 nm (**Figure 5G**) using SDS aqueous

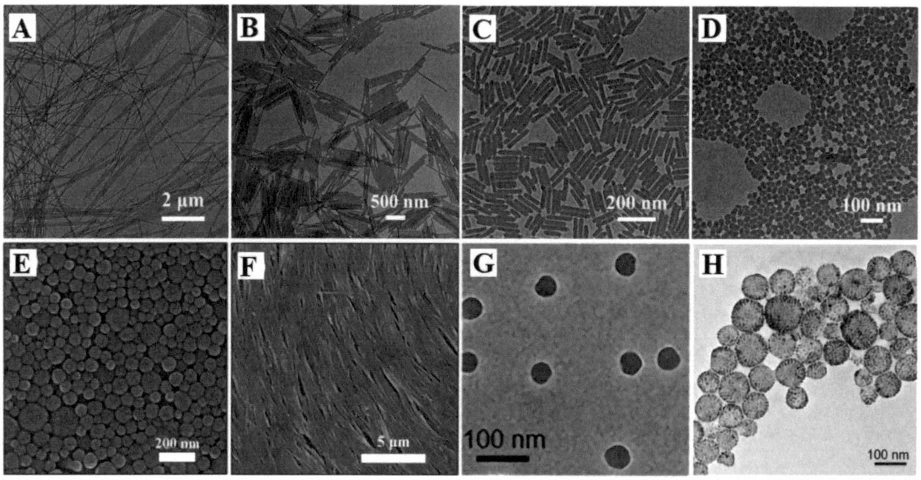

Figure 5.
Structure characterizations of the self-assembled porphyrin nanostructures by microemulsion-assisted/mesophase transfer self-assembly method. TEM images of the self-assembled InTPP nanocrystals at different CTAB concentrations of (A) 2.5, (B) 10, (C) 15, and (D) 25 mM, respectively. Reproduced with permission from reference [16]. Copyright 2019 American Chemical Society. SEM images of the self-assembled of MnPc NPs (E) and MnPc NWs (F). Reproduced with permission from reference [17]. Copyright 2021 Tsinghua University Press and Springer-Verlag GmbH Germany, part of Springer Nature 2021. TEM images of the self-assembled of AuTPyP NPs (G) and FeTPP@Fe_3O_4 (H). (G) Reproduced with permission from reference [18]. Copyright 2021 American Chemical Society. (H) Reproduced with permission from reference [19]. Copyright 2022 Tsinghua University Press and Springer-Verlag GmbH Germany, part of Springer Nature 2022.

solution/$CHCl_3$ as microemulsion droplet [18]. More interestingly, the microemulsion droplet technology can well integrate the organic–inorganic interface, and then be used to construct the organic–inorganic composite coassemblies. Typically, the $CHCl_3$ solution of 5,10,15,20-tetraphenyl-21H, 23H-porphine iron (III) chloride (FeTPP) and oleic acid (OA) modified magnetic Fe_3O_4 NCs was injected into SDS aqueous solution to create an oil-in-water microemulsion by an ultrasonic probe [19]. After evaporation of the chloroform solvent, the FeTPP and OA-Fe_3O_4 NCs randomly coassembled into nanospheres driven by π-π stacking and hydrophobic-hydrophobic interactions (**Figure 5H**).

3. Photoelectric response of porphyrin self-assemblies

Optical spectroscopy has been used to study the light absorption properties of porphyrin nanostructures [20, 21]. The spectra provide information about the intermolecular interactions in porphyrins nanostructures [22, 23]. The UV–vis absorption spectra of self-assembled ZnTPyP with different morphologies show that the porphyrin nanostructure has a larger spectral absorption broadening and obvious redshift compared with the porphyrin monomer. After self-assembled into the nanostructures (the nanodisk, hexagonal nanorod, tetragonal nanorod, and nanoparticle), the absorption at 425 nm of the Soret band porphyrin monomer is split into three absorption peaks at 421 nm, 449 nm, and 475 nm, respectively. The tetragonal nanorod absorbs at 421 nm and 448 nm, and nanoparticle absorbs at 421 nm. The nanodisk has the largest redshift, which is usually associated with the spatially ordered J-aggregate arrangement of porphyrin molecules. Basically, ordered π-π stacking and long-range delocalization within porphyrin assemblies enable these nanomaterials with tailored collective optical properties for a broader visible light

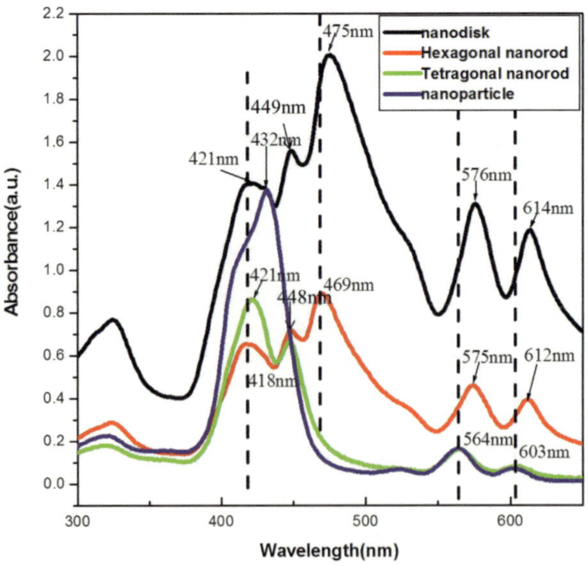

Figure 6.
UV–vis absorptions of zinc-tetra(4-pyridyl) porphyrin (ZnTPyP) self-assembled nanocrystals. Reproduced with permission from reference [20]. Copyright 2014 American Chemical Society.

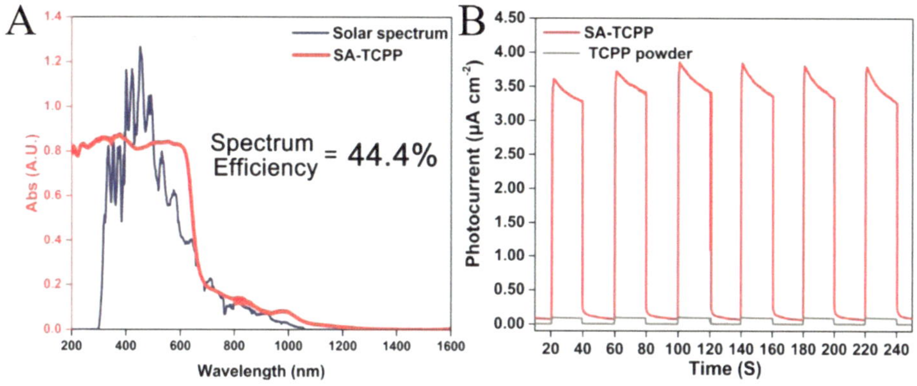

Figure 7.
(A) UV-vis diffuse reflection spectroscopy of TCPP self-assembled nanostructures. (B) Photocurrent response of SA-TCPP and TCPP powder. Reproduced with permission from reference [24]. Copyright 2018 Wiley-VCH.

absorption spectrum. Since the pyridyl group in the ZnTPyP molecule has a certain angle with the plane of the porphyrin ring, the intermolecular force of the porphyrin molecule in the self-assembly process is different, which lead to different close-packed stacking and different morphologies [20]. The relationship between visible absorption spectra-nanostructures of porphyrin nanostructures will also seriously affect the photocatalytic activity (**Figure 6**).

The self-assembled tetra(4-carboxylphenyl) porphyrin (TCPP) (SA-TCPP) exhibits a single absorption edge at 700 nm due to the spatial order aggregation of TCPP molecules, suggesting the SA-TCPP could absorb the entire visible spectrum (**Figure 7A**). In addition, the SA-TCPP could further utilize the UV spectrum because of the blue-shift of the Soret band. Therefore, the SA-TCPP could collect the full solar spectrum light and their theoretical spectral efficiency reached up to 44.4% [24]. Compared with the commercial TCPP powder, the photocurrent of SA-TCPP is as high as 3.52 μA cm^{-2} (**Figure 7B**), further confirming that the ordered packing of porphyrin molecules is conducive to electron delocalization and transition and further enhances the photoelectric response performance.

4. Applications of porphyrin self-assemblies

4.1 Photocatalytic degradation of methyl orange (MO)

Based on the excellent light-harvesting ability of porphyrins [25], porphyrin self-assembled nanostructures are used in the field of photocatalysis. H_2TPyP nanooctahedra, Zn-metallized H_2TPyP intermediate nanoparticles, and Zn-metallized H_2TPyP nanowires were used for photocatalytic degradation of methyl orange (MO) [26]. As shown in **Figure 8**, the regular 200 nm H_2TPyP nanooctahedra has a degradation efficiency of about 10% after 140 minutes of illumination (panel b). The degradation efficiency of the mixture of octahedra and long wires after part of Zn-metallized, degradation efficiency increased to nearly 80% (panel c), and the equivalent degradation efficiency of Zn-metallized completely porphyrin long wires is close to 90% (panel d). The self-degradation of MO as a control was almost unchanged under the same condition [11]. These results confirmed that the addition of center metal atoms

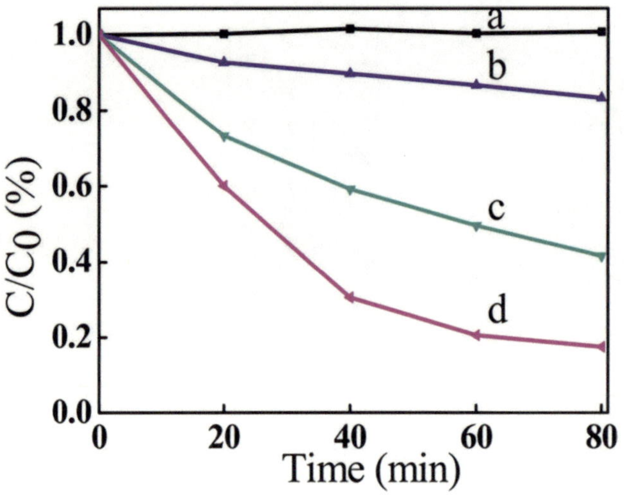

Figure 8.
Photocatalytic degradation of methyl orange (MO). (a) Self-degradation of MO as a control. (b) H_2TPyP nanooctahedra. (c) Zn-metallized H_2TPyP intermediate nanoparticles. (d) Zn-metallized H_2TPyP nanowires. Reproduced with permission from reference [11]. Copyright 2016 American Chemical Society.

played a very important role in the improvement of the catalytic efficiency. On the other hand, the long-range ordered π-π conjugated stacking in nanowires promoted better electron transport and electron–hole separation, thereby showing better photocatalytic degradation efficiency than that of nanooctahedra. In addition, porphyrin nanostructures have the advantages of less dosage, high activity, easy recycling, and stability in photocatalytic applications, which makes them a potential photocatalytic agent.

4.2 Photocatalyzed hydrogen evolution

The molecular ordered self-assembled PdTCPP nanoleaf has strong absorption in the range of 400 to 700 nm and high photogenerated charge separation efficiency under visible light irradiation. In the absence of cocatalyst, the hydrogen production rate of PdTCPP nanoleaf (9.9 mmol g^{-1} h^{-1}) was nearly 40 times higher than that of commercial PdTCPP powder (0.25 mmol g^{-1} h^{-1}) [12]. After loading 1 wt% Pt cocatalyst, the hydrogen production rate of PdTCPP nanoleaf reached 138 mmol g^{-1} h^{-1}, which was 14 times higher than that without cocatalyst (**Figure 9A**). This spatial arrangement of porphyrin molecules' self-assembly effectively improved electron transport to enhance photocatalytic hydrogen production. Interestingly, TEM images showed that Pt nanoparticles were mainly deposited on the edge of the PdTCPP nanoleaf after the photocatalytic reaction, while there were almost no Pt nanoparticles on the surface when Pt was cocatalyst (**Figure 9B**). That is, the edge of the PdTCPP nanoleaf is the active center for the photocatalytic reduction reaction. When K_2PtCl_4 and $Pb(NO_3)_2$ were used as precursors to conduct photodeposition experiments, TEM images showed that Pt and PbO_2 nanoparticles were deposited on the edge and surface of PdTCPP nanoleaf, respectively (**Figure 9C**). The element distribution maps show that Pt and Pb elements are concentrated on the edge and surface, respectively, that is, Pt nanoparticles are selectively deposited on the edge of PdTCPP nanoleaf, while PbO_2 nanoparticles tend to be deposited on the surface of PdTCPP nanoleaf (**Figure 9D–G**). The above

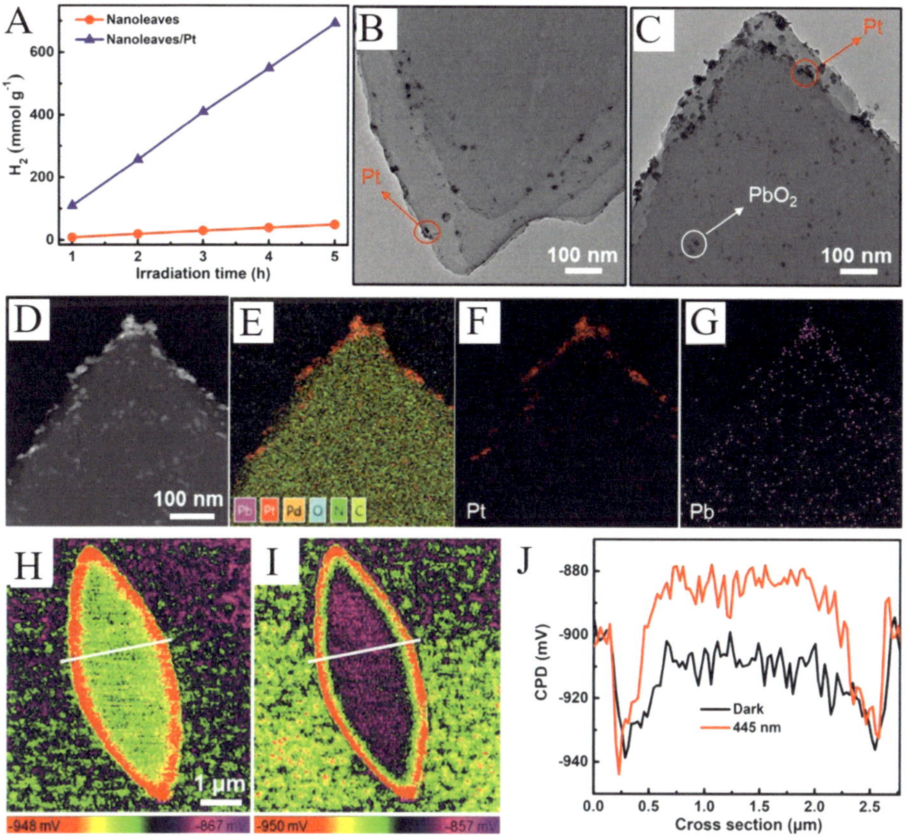

Figure 9.
(A) H_2 production photocatalyzed by PdTCPP nanoleaves without and with 1 wt % of Pt cocatalyst. (B) TEM image of the nanoleaves with Pt NPs. (C) TEM image of the nanoleaves with Pt and PbO_2 NPs. (D–G) elements mapping of the nanoleaves/Pt/PbO_2. (H) under dark (I) and light irradiation at 445 nm of kelvin probe force microscopy images of the nanoleaves. (J) surface potential image of the cross-section in (H) and (I). Reproduced with permission from reference [12]. Copyright 2022 American Chemical Society.

experimental results indicated that photogenerated electrons and holes are separated and transported in different directions in the PdTCPP nanoleaf. Kelvin probe force microscopy (KPFM) can provide the contact potential difference (CPD) between the tip and the sample, which can detect the charge distribution inside the single crystal. The KPFM surface potential of PdTCPP nanoleaf shown that in the dark state, the edge potential of PdTCPP nanoleaf is more negative than the surface, and there is a potential difference between edge and surface. The surface potential of PdTCPP nanoleaf becomes more positive and significantly increased under 445 nm laser irradiation, the potential difference was significantly enhanced, confirming the existence of space charge regions between the edges and faces of the PdTCPP nanoleaf (**Figure 9H–J**). This work reported for the first time there is an electric field inside the PdTCPP nanoleaf, and the photogenerated electrons and holes were transferred and gathered to the surface under visible light irradiation, respectively. This selective transfer of photogenerated electrons and holes to specific sites on the edge and surface of PdTCPP nanoleaf immensely improved the spatial separation efficiency of photogenerated charges, thereby enhancing photocatalytic activity.

4.3 Photodynamic therapy

Porphyrins as a class of biocompatible and easily modified photosensitizers, have shown great application value in tumor photodynamic therapy (PDT) [27–30]. The co-assembly of biomimetic Gd(III) meso-tetraphenyl porphyrin 2,4-pentane dionate (GdTPP) and zinc meso-tetraphenyl porphyrin (ZnTPP) mGZNs were used for PDT functions research of HeLa cells [31]. Firstly, singlet oxygen ($^1O_2$) has been shown to be the key photooxidation species for PDT that efficiently kills cancer cells [32]. The PDT efficiency and the generation of intracellular reactive oxygen species (ROS) were further evaluated by the specific green fluorescence probe (2′,7′-dichlorodihydrofluorescein diacetate, DCFH-DA). The comparison of PBS control was performed and showed mGNs have the highest fluorescence intensity of ROS including $^1O_2$ for PDT (**Figure 10A**). Additionally, a clear, nearly linear increase in fluorescence intensity as mGZNs concentration increased was shown (**Figure 10B**). Cell viability indicated an irradiation time- and dosage-dependent PDT efficiency with a maximum value of 80.6% at 200 μM of mGZNs and an irradiation duration of 9 min (635 nm, 0.15 W/cm^2; **Figure 10C**). Moreover, flow cytometry analyses of HeLa necrosis and apoptosis induced by mGZNs showed that irradiation of mGZNs induced greater cell death compared to the blank groups (**Figure 10D**).

4.4 Synergistic chemo-photothermal therapy

The random aggregation of porphyrin molecule will limit the separation and transfer of photoelectrons and charges, which is more conducive to photothermal conversion. AuPNS showed a dose- and irradiation time-dependent photothermal effect. After 10 min of irradiation, the AuPNS solution rapidly increased to 78°C, and the photothermal conversion efficiency of AuPNS was ~48.2% (**Figure 11A**) [18]. The infrared thermal image of the AuPNSs droplets can reach up to 50.2°C, while the deionized water droplets hardly change as control, and the AuPNSs show excellent photothermal stability without obvious photothermal decay after five irradiation cycles (**Figure 11B**). Water-insoluble pyridyl porphyrins can be protonated into water-soluble monomers under acidic conditions. Moreover, high temperatures can accelerate the release of pyridyl porphyrin in an acidic solution. Therefore, the release percentage of AuPNS showed that AuPNS was rapidly released at high temperature (50°C) and low pH (5.0) with a release rate as high as 54.0% (**Figure 11C**). Subsequently, AuPNS was applied to the antitumor activity test. After modifying cRGD-AuPNS with its cRGD, the confocal images showed stronger red fluorescence, indicating that cRGD promoted the endocytosis of nanodrugs (**Figure 11D**). Time-dependent flow cytometry analysis showed that cRGD-AuPNS reached a maximum after 6 hours of incubation in the cytoplasm (**Figure 11E** and **F**). Au(III) in AuTPyP molecule can inhibit the activity of TrxR and enhance ROS production (**Figure 11G**). Quantitative testing of the kit revealed potent TrxR inhibition and ROS generation (**Figure 11H**). Since the level of TrxR in tumor cells was higher than that in normal cells, cRGD-AuPNS showed chemotherapy-killing effect on tumor cells, but no obvious killing on normal 293 T cells. In particular, the killing of HeLa cells was the most pronounced and reached to 85.2% (**Figure 11I**).

4.5 Peroxidase-like activity for anti-tumor therapy

Porphyrins and phthalocyanines (Pc) are a class of N-doped carbon-based macrocyclic compounds, and the central ring can combine with different metal ions

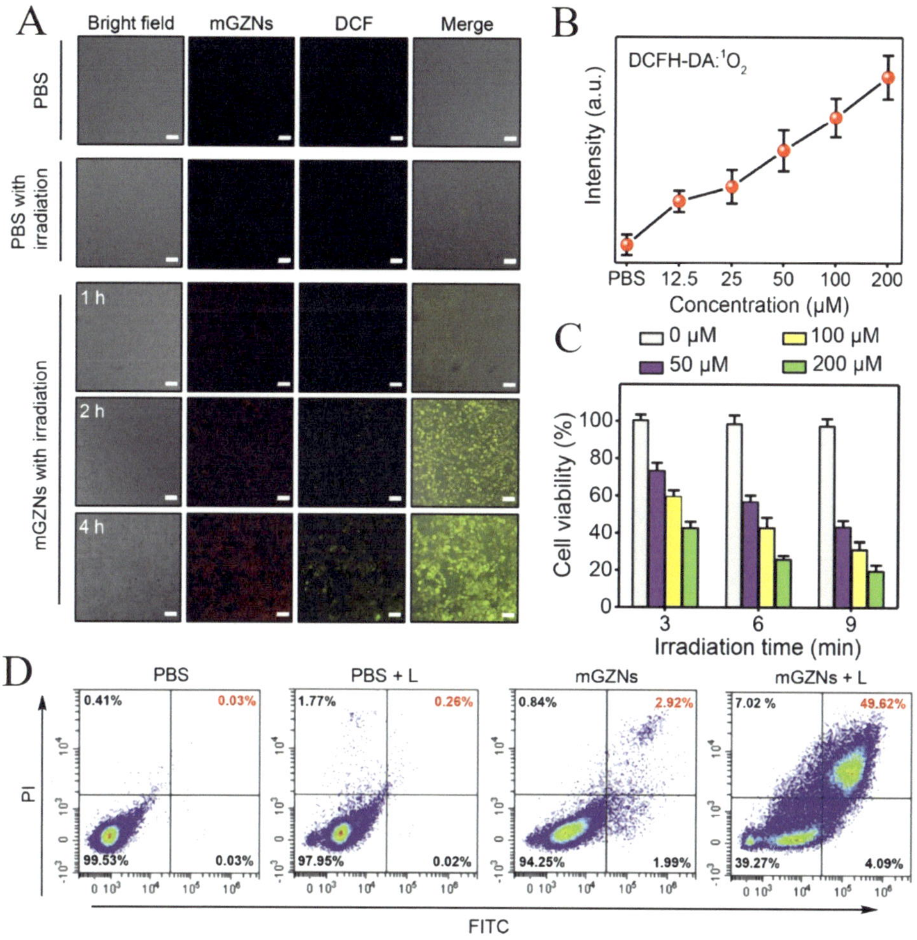

Figure 10.
In vitro experiments with mGZNs. (A) Images showing intracellular production of various porphyrin self-assembled nanostructures. Scale bar: 100 μm. (B) the fluorescence intensity of $^1O_2$ in cells. (C) Irradiation durations and dosages-dependent phototoxicity of HeLa cells incubated with porphyrin self-assembled nanostructures. (D) Flow cytometry analyses of annexin V-FITC/PI double-stained cells treated with various porphyrin self-assembled nanostructures. Reproduced with permission from reference [31]. Copyright 2020 Elsevier.

to form metal complexes. Among them, iron porphyrins are often used to simulate the catalysis of biomimetic enzymes *in vivo* [33]. Similar to porphyrins, Pc has been widely used in biological phototherapy. Manganese phthalocyanine (MnPc) molecules coordinated with divalent manganese ions (Mn^{2+}) were selected as the building blocks, and MnPc assemblies were prepared by microemulsion droplets method. The blended assembly of emulsifier and MnPc reduced the close-packed structure of MnPc molecules, exposed more active sites, and exhibited morphology-dependent POD-like catalytic activity [17]. The kinetic curve of POD-like catalytic reaction of MnPc nanoparticles (MnPcNPs), MnPc nanowires (MnPcNWs), tetraphenyl manganese (III) porphyrin (MnTPP) assemblies (**Figure 12A**), tetrapyridyl manganese (III) porphyrin (MnTPyP) and N, N-diethyl-4-anilinoporphyrin (MnTNPP) single molecule shows that MnPcNPs exhibit the best POD-like catalytic performance, while MnTPPNPs have almost no catalytic activity (**Figure 12B**). The catalytic

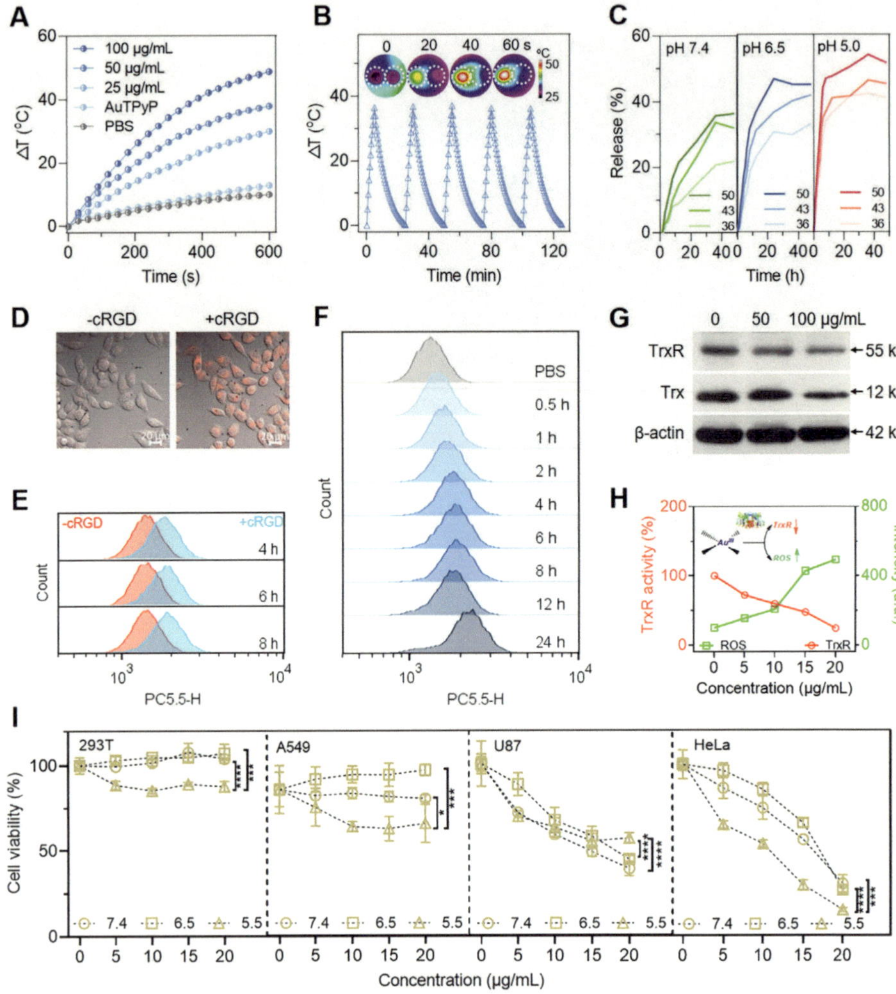

Figure 11.
Synergistic chemo-photothermal therapy of AuTPyP self-assembled nanospheres (AuPNSs). (A) AuPNSs concentrations-dependent photothermal effect. (B) Photothermal stability of AuPNSs. (C) pH values and temperatures stimuli-responsive released plot profile of AuPNSs. (D) Confocal images of HeLa cells treated by cRGD modified AuPNSs. (E) the corresponding flow cytometry analyses of (D). (F) the detailed time-dependent uptake in cRGD-AuPNSs group. (G) Western blot analyses of TrxR and Trx proteins in HeLa cells treated with cRGD-AuPNSs. (H) Kits test of (G). (I) Cytotoxicity comparison of various cell lines treated with cRGD-AuPNSs. Reproduced with permission from reference [18]. Copyright 2021 American Chemical Society.

performance of MnTPyP molecules is better than that of MnTNPP molecules, and this difference in catalytic performance is more intuitively reflected in the color change of the reaction system (**Figure 12C**). These results indicate that the catalytic activity of MnPc assemblies is higher than that of Mn porphyrin assemblies, which may be related to the larger N doping ratio in the phthalocyanine ring and the different oxidation states exhibited by the central metal. MnPc assemblies can be used as a nanozyme to simulate the POD catalytic performance, and exhibits morphology, pH value, and temperature-dependent catalytic performance [34]. The nanomaterials are expected to realize catalytic properties in tumor microenvironment for antitumor therapy.

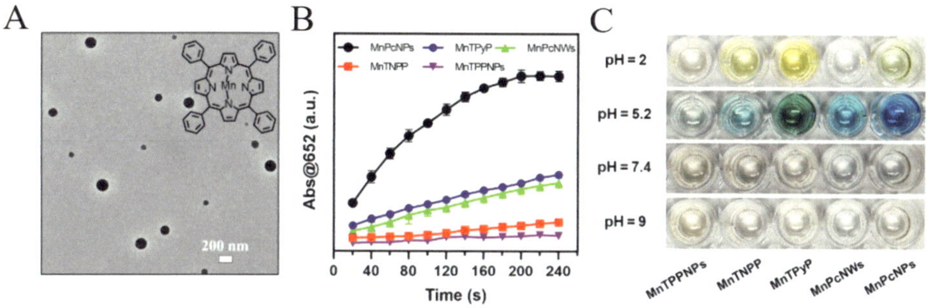

Figure 12.
(A) TEM of MnTPPNPs. (B) the time-dependent absorbance change curves of various porphyrin-assembled nanostructures. (C) Photographs of TMB peroxidation catalyzed by different materials. Reproduced with permission from reference [17]. Copyright 2021 Tsinghua University Press and Springer-Verlag GmbH Germany, part of Springer Nature 2021.

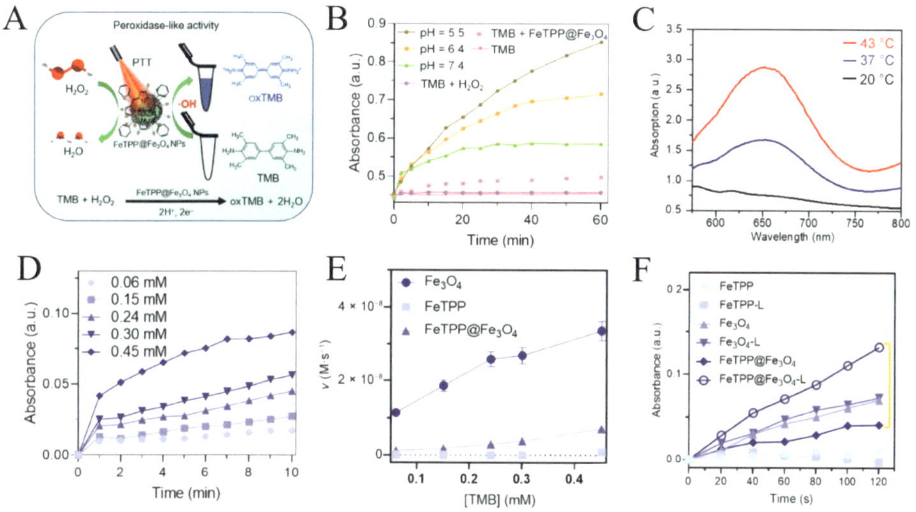

Figure 13.
Photo-enhanced peroxidase-like activity of FeTPP@Fe_3O_4 NPs. (A) Schematic illustration of photo-enhanced POD-like activity of FeTPP@Fe_3O_4 NPs. (B) the pH-dependent and (C) temperatures-dependent POD-like activity of FeTPP@Fe_3O_4 NPs. (D) the time-course POD-like activity of FeTPP@Fe_3O_4 NPs as a function of concentration of TMB. (E) the corresponding Michaelis–Menten kinetics and (F) photo-enhanced POD-like activity of various nanoparticles. Reproduced with permission from reference [19]. Copyright 2022 Tsinghua University Press and Springer-Verlag GmbH Germany, part of Springer Nature 2022.

Since the POD catalytic performance exhibits temperature-dependent catalytic performance. Similarly, Fe_3O_4 NCs have excellent POD properties (**Figure 13A**), but due to the limitation of insufficient hydrogen peroxide (H_2O_2) concentration in tumor cells, the generated hydroxyl radicals (•OH) from Fe_3O_4 Fenton reaction are far from enough to kill tumor cells [35, 36]. Through the self-assembly driving force of porphyrins, FeTPP@Fe_3O_4 NPs were prepared by microemulsion droplets method. Under laser irradiation, the photothermal generated by FeTPP assemblies promotes POD performance of Fe_3O_4 NPs, and instantaneously generates a large number of free hydroxyl groups to kill tumor cells (**Figure 13B–F**). These results clearly demonstrate that FeTPP@Fe_3O_4 NPs can effectively accelerate the catalytic decomposition of endogenous H_2O_2 in an acidic environment under laser irradiation, further accelerate the generation of •OH, and improve the CDT efficacy *in vivo* [19].

5. Conclusions

Porphyrins possess excellent light-harvesting properties, and have tunable spectral features, functional groups, and central metals, which endow them with unique photophysical properties. Porphyrin self-assembled nanostructures will inherit the characteristics of porphyrin monomer, which make them widely used in photocatalysis, tumor therapy, and other fields. Porphyrin nanostructures have the advantages of less dosage, high activity, easy recycling, stability in photocatalytic applications, excellent photothermal and photodynamic therapy, and nanozyme catalytic performance, which makes them a potential photocatalytic agent and nanotherapeutic platform. However, the current research on the assembly mechanism is still unclear. Various in situ techniques and integrating-sphere-assisted resonance synchronous (ISARS) spectroscopy methods need to be introduced to detect the self-assembly process for a deeper understanding of the packing nucleation process [37]. Moreover, the influence of external fields on the assembly process of porphyrin molecules, such as temperature, humidity, magnetic field, electric field, ultrasound, high-speed centrifugation, need to be urgently explored in further works. More importantly, the structure–activity relationship between the porphyrin morphology structure and performance will provide important technical support and experimental basis for revealing the self-assembly process in living organisms. The exploration of the self-assembly of porphyrins with biofunctional molecules, such as amino acids, polypeptides, and nucleic acids, are very intriguing research direction to form a mature light harvest antenna [38] and self-assembled nanomedicine platform [39, 40], which may achieve major breakthroughs in life sciences. This chapter will provide an important reference for the cognition of self-assembled nanostructures.

Acknowledgements

This work was supported by the National Natural Science Foundation of China (21802032 and U21A2085), China Postdoctoral Science Foundation (2019TQ0081), and Zhongyuan high-level talents special support plan (No. 204200510009).

Conflict of interest

The authors declare no conflict of interest.

Author details

Shiqi Liu†, Zengyin Li†, Huiwei Tong, Yong Zhong* and Feng Bai*
Key Laboratory for Special Functional Materials of Ministry of Education, National and Local Joint Engineering Research Center for High-efficiency Display and Lighting Technology, School of Materials Science and Engineering, Collaborative Innovation Center of Nano Functional Materials and Applications, Henan University, Kaifeng, P.R. China

*Address all correspondence to: zhywy8521115@163.com and baifengsun@126.com

† Shiqi Liu and Zengyin Li contributed equally to this work.

References

[1] Wei W, Sun J, Fan H. Cooperative self-assembly of porphyrins and derivatives. MRS Bulletin. 2019;**44**:178-182. DOI: 10.1557/mrs.2019.39

[2] Huang C, Li Y, Song Y, Li Y, Liu H, Zhu D. Ordered nanosphere alignment of porphyrin for the improvement of nonlinear optical properties. Advanced Materials. 2010;**22**:3532-3536. DOI: 10.1002/adma.200904421

[3] Sandanayaka ASD, Arakiet Y, Wada T, Hasobe T. Structural and photophysical properties of self-assembled porphyrin nanoassemblies organized by ethylene glycol derivatives. Journal of Physical Chemistry C. 2008;**112**:19209-19216. DOI: 10.1021/jp805202y

[4] Ji W, Xue B, Bera S, Guerin S, Liu Y, Yuan H, et al. Tunable mechanical and optoelectronic properties of organic cocrystals by unexpected stacking transformation from H- to J- and X-aggregation. ACS Nano. 2020;**14**:10704-10715. DOI: 0.1021/acsnano.0c05367

[5] Huang C, Wen L, Liu H, Li Y, Liu X, Yuan M, et al. Controllable growth of 0D to multidimensional nanostructures of a novel porphyrin molecule. Advanced Materials. 2009;**21**:1721-1725. DOI: 10.1002/adma.200802114

[6] van Hameren R, Schon P, van Buul AM, Hoogboom J, Lazarenko SV, Gerritsen JW, et al. Macroscopic hierarchical surface patterning of porphyrin trimers via self-assembly and dewetting. Science. 2006;**314**:1433-1436. DOI: 10.1126/science.1133004

[7] Martin KE, Wang Z, Busani T, et al. Donor–acceptor biomorphs from the ionic self-assembly of porphyrins. Journal of the American Chemical Society. 2010;**132**:8194-8201. DOI: 10.1021/ja102194x

[8] Zhang N, Wang L, Wang H, Cao R, Wang J, Bai F, et al. Self-assembled one-dimensional porphyrin nanostructures with enhanced photocatalytic hydrogen generation. Nano Letters. 2018;**18**(1):560-566. DOI: 10.1021/acs.nanolett.7b04701

[9] Zhong Y, Hu Y, Wang J, Wang J, Ren X, Sun J, et al. Morphology and size-dependent visible-light-driven photocatalytic hydrogen evolution of porphyrin assemblies. MRS Advances. 2019;**4**:2071-2078. DOI: 10.1557/adv.2019.210

[10] Chen S, Ren X, Tian S, Sun J, Bai F. Controllable synthesis of cobalt porphyrin nanocrystals through micelle confinement self-assembly. MRS Advances. 2020;**5**(42):2147-2155. DOI: 10.1557/adv.2020.269

[11] Wang J, Zhong Y, Wang L, Zhang N, Cao R, Bian K, et al. Morphology-controlled synthesis and metalation of porphyrin nanoparticles with enhanced photocatalytic performance. Nano Letters. 2016;**16**(10):6523-6528. DOI: 10.1021/acs.nanolett.6b03135

[12] Cao R, Wang G, Ren X, Duan PC, Wang L, Li Y, et al. Self-assembled porphyrin nanoleaves with unique crossed transportation of photogenerated carriers to enhance photocatalytic hydrogen production. Nano Letters. 2022;**22**(1):157-163. DOI: 10.1021/acs.nanolett.1c03550

[13] Qiu Y, Chen P, Liu M. Evolution of various porphyrin nanostructures via

an oil/aqueous medium: Controlled self-assembly, further organization, and supramolecular chirality. Journal of the American Chemical Society. 2010;**132**(28):9644-9652. DOI: 10.1021/ja1001967

[14] Wei W, Bai F, Fan H. Surfactant-assisted cooperative self-assembly of nanoparticles into active nanostructures. iScience. 2019;**11**:272-293. DOI: 10.1016/j.isci.2018.12.025

[15] Zhong Y, Wang Z, Zhang R, Bai F, Wu H, Haddad R, et al. Interfacial self-assembly driven formation of hierarchically structured nanocrystals with photocatalytic activity. ACS Nano. 2014;**8**(1):827-833. DOI: 10.1021/nn405492d

[16] Liu Y, Wang L, Feng H, Ren X, Ji J, Bai F, et al. Microemulsion-assisted self-assembly and synthesis of size-controlled porphyrin nanocrystals with enhanced photocatalytic hydrogen evolution. Nano Letters. 2019;**19**(4):2614-2619. DOI: 10.1021/acs.nanolett.9b00423

[17] Wang J, Gao S, Wang X, Zhang H, Ren X, Liu J, et al. Self-assembled manganese phthalocyanine nanoparticles with enhanced peroxidase-like activity for anti-tumor therapy. Nano Research. 2021;**15**(3):2347-2354. DOI: 10.1007/s12274-021-3854-5

[18] Wang X, Wang J, Wang J, Zhong Y, Han L, Yan J, et al. Noncovalent self-assembled smart gold (III) porphyrin nanodrug for synergistic chemo-photothermal therapy. Nano Letters. 2021;**21**(8):3418-3425. DOI: 10.1021/acs.nanolett.0c04915

[19] Tian T, Bao J, Wang J, Wang J, Ge Y, Li Z, et al. Co-assembly of FeTPP@Fe_3O_4 nanoparticles with photo-enhanced catalytic activity for synergistic tumor therapy. Nano Research. 2022. DOI: 10.1007/s12274-022-4548-3

[20] Zhong Y, Wang J, Zhang R, Wei W, Wang H, Lu X, et al. Morphology-controlled self-assembly and synthesis of photocatalytic nanocrystals. Nano Letters. 2014;**14**(12):7175-7179. DOI: 10.1021/nl503761y

[21] Zhao Y, Hu Y, Zhong Y, Wang J, Liu Z, Bai F, et al. Missing links between the structures and optical properties of porphyrin assemblies. Journal of Physical Chemistry C. 2021;**125**(40):22318-22327. DOI: 10.1021/acs.jpcc.1c06795

[22] Wang L, Fan H, Bai F. Porphyrin-based photocatalysts for hydrogen production. MRS Bulletin. 2020;**45**(1):49-56. DOI: 10.1557/mrs.2019.294

[23] Li Q, Bao Y, Bai F. Porphyrin and macrocycle derivatives for electrochemical water splitting. MRS Bulletin. 2020;**45**(7):569-573. DOI: 10.1557/mrs.2020.168

[24] Zhang Z, Zhu Y, Chen X, Zhang H, Wang J. A full-spectrum metal-free porphyrin supramolecular photocatalyst for dual functions of highly efficient hydrogen and oxygen evolution. Advanced Materials. 2019;**31**(7):e1806626. DOI: 10.1002/adma.201806626

[25] Li Q, Zhao N, Bai F. Size- and shape-dependent photocatalysis of porphyrin nanocrystals. MRS Bulletin. 2019;**44**(3):172-177. DOI: 10.1557/mrs.2019.41

[26] Zhong Y, Liu S, Wang J, Zhang W, Tian T, Sun J, et al. Self-assembled supramolecular nanostructure photosensitizers for photocatalytic hydrogen evolution. APL Materials. 2020;**8**(12):120706. DOI: 10.1063/5.0029923

[27] Wang D, Niu L, Qiao ZY, Cheng DB, Wang J, Zhong Y, et al. Synthesis of

self-assembled porphyrin nanoparticle photosensitizers. ACS Nano. 2018;**12**(4):3796-3803. DOI: 10.1021/acsnano.8b01010

[28] Wang D, Cheng DB, Ji L, Niu LJ, Zhang XH, Cong Y, et al. Precise magnetic resonance imaging-guided sonodynamic therapy for drug-resistant bacterial deep infection. Biomaterials. 2021;**264**:120386. DOI: 10.1016/j.biomaterials.2020.120386

[29] Wang J, Hu Y, Wang X, Gao S, Zhong Y, Liu J, et al. Trace-water-induced competitive coordination synthesis and functionalization of porphyrinic metal-organic framework nanoparticles for treatment of hypoxic tumors. ACS Applied Bio Materials. 2021;**4**(9):7322-7331. DOI: 10.1021/acsabm.1c00852

[30] Shi Y, Liu S, Liu Y, Sun C, Chang M, Zhao X, et al. Facile fabrication of nanoscale porphyrinic covalent organic polymers for combined photodynamic and photothermal cancer therapy. ACS Applied Materials & Interfaces. 2019;**11**(13):12321-12326. DOI: 10.1021/acsami.9b00361

[31] Wang J, Wang Z, Zhong Y, Zou Y, Wang C, Wu H, et al. Central metal-derived co-assembly of biomimetic GdTPP/ZnTPP porphyrin nanocomposites for enhanced dual-modal imaging-guided photodynamic therapy. Biomaterials. 2020;**229**:119576. DOI: 10.1016/j.biomaterials.2019.119576

[32] Zhang H, Pan X, Wu Q, Guo J, Wang C, Liu H. Manganese carbonate nanoparticles-mediated mitochondrial dysfunction for enhanced sonodynamic therapy. Exploration. 2021;**1**(2):20210010. DOI: 10.1002/EXP.20210010

[33] Monti D, Venanzi M, Cantonetti V, Borocci S, Mancini G. Effect of surfactant phase transition on the inclusion behaviour of an amphiphilised porphyrin derivative. Chemical Communications. 2002;**7**:774-775. DOI: 10.1039/B111692D

[34] Jana D, Zhao Y. Strategies for enhancing cancer chemodynamic therapy performance. Exploration. 2022;**2**(2):20210238. DOI: 10.1002/EXP.20210238

[35] Cui X, Zhang Z, Yang Y, Li S, Lee CS. Organic radical materials in biomedical applications: State of the art and perspectives. Exploration. 2022;**2**(2):20210264. DOI: 10.1002/EXP.20210264

[36] Tang G, He J, Liu J, Yan X, Fan K. Nanozyme for tumor therapy: Surface modification matters. Exploration. 2021;**1**(1):75-89. DOI: 10.1002/EXP.20210005

[37] Wamsley M, Wathudura P, Hu J, Zhang D. Integrating-sphere-assisted resonance synchronous spectroscopy for the quantification of material double-beam uv–vis absorption and scattering extinction. Analytical Chemistry. 2022;**94**(33):11610-11618. DOI: 10.1021/acs.analchem.2c02037

[38] Liu K, Xing R, Li Y, Zou Q, Möhwald H, Yan X. Mimicking primitive photobacteria: Sustainable hydrogen evolution based on peptide-porphyrin co-assemblies with a self-mineralized reaction center. Angewandte Chemie, International Edition. 2016;**55**:12503-12507. DOI: 10.1002/anie.201606795

[39] Mamuti M, Zheng R, An H, Wang H. *In vivo* self-assembled nanomedicine. Nano Today. 2021;**36**:101036. DOI: 10.1016/j.nantod.2020.101036

[40] Cai J, Liu X, Gao Z, Li L, Wang H. Chlorophylls derivatives: Photophysical properties, assemblies, nanostructures and biomedical applications. Materials Today. 2021;**45**:77-92. DOI: 10.1016/j.mattod.2020.11.001

Chapter 3

Amphiphiles

Beheshteh Sohrabi

Abstract

Amphiphiles form a large group of supramolecular structures can aggregate and be adsorbed spontaneously at the interface. Amphiphilicity is a feature of polar contrast between the groups that make up a molecule and their spatial separation. The most important classes of amphiphiles are surfactants, lipoproteins, and polymers that have hydrophilic and hydrophobic chemical moieties covalently bonded and spatially separated. Since surfactants are widely used in various industrial fields, we decide to focus on surfactants in addition to a brief review of the other amphiphiles. Surfactants are used in industrial applications and consumer products, from medical to cosmetics and food industry. Various industries require new surfactants from sustainable and renewable raw materials with improved performance, biocompatibility and minimal environmental impact. For example, liquid phase exfoliation and dispersion methods using surfactants in the solvent media have recently gained lots of attention because of their great potential for large-scale production. Notably, an ideal exfoliation for reaching desired graphene and CNTs may be achievable by molecular engineering of surfactants to improve the quality of molecular interactions. This chapter experimentally and theoretically highlighted physico-chemical characteristic parameters, and interactions of the components, which are essential to design and discover efficient exfoliation and dispersion systems.

Keywords: amphiphilic, conventional amphiphiles, advanced amphiphiles, hdyrophilicity, hydrophobicity, surfactants, critical micelle concentration

1. Introduction

Since amphiphiles mainly form a large group of supramolecular structures, it is important to study their structure and properties. Amphiphilicity is one of the important structural features because two opposing parts (hydrophilic and hydrophobic) exist simultaneously in the structure of a molecule. This property causes such molecules to be able to spontaneously assemble at the interface of polar and non-polar two phases or to be forced to self-assemble to be in one phase. According to **Figure 1**, amphiphiles are molecules consisting of a hydrophilic "head" and a hydrophobic "tail", which the two parts are linked on the basis of covalent bonds [1]. As can see in **Figure 2**, their hydrophilic heads can contain ionic, polar, non-ionic groups or form hydrogen bonds that lead to their dissolution in water. Their hydrophobic tail may

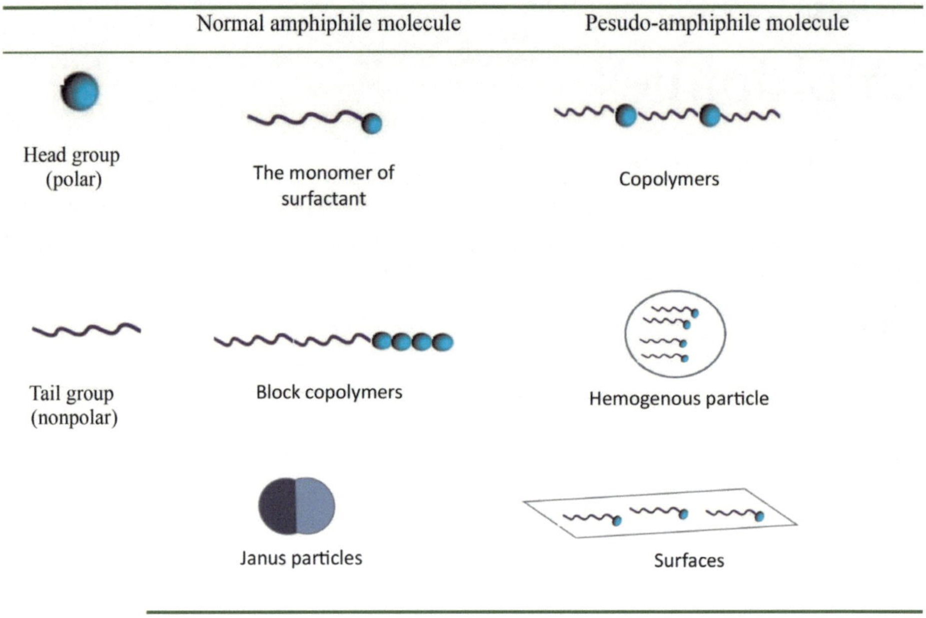

Figure 1.
The structure of amphiphiles and pseudo-amphiphiles.

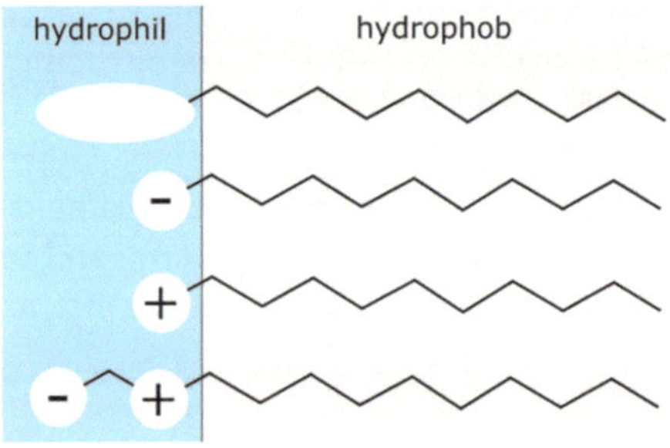

Figure 2.
Schematic structure of amphiphiles.

also include saturated or unsaturated hydrocarbon chains or cyclic hydrocarbons. Their dual nature leads to spontaneous placement at the boundary between two phases, such as liquid–liquid, liquid–gas or solid–liquid, as a result reducing the surface tension and surface energy between the phases. Amphiphilicity is a scalable property that occurs in a molecular-level, nanoscale system and can probably be seen on a micro scale. The scalability of amphiphilic properties is a fundamentally new and hot topic and is considered in basic research [2–4].

2. Brief historical account of amphiphiles

The word amphiphile comes from the Greek words amphis, meaning "both," and philia, meaning "love." Surfactants are found as amphiphiles in plants such as tomatoes, onions, mulberry leaves, ginseng, licorice, and all plants that contain glucosides. In the past, humans used mulberry leaves to clean their hands, which were stained with blackberry juice, without being aware of the dual nature of these substances and their washing.

Soap is another amphiphilic substance that has been made and used since ancient times in Mesopotamia [5]. Soap is a surfactant of fatty acids, and in the past it was made by natural synthetic methods from natural raw materials such as tallow, olive, argan or palm oil, and alkali-rich ash, the ash left over from wood burning. After the preparation of soap in Mesopotamia, its production has spread throughout continental Europe (**Figure 3**). In addition to helping to improve the quality of life, soap can eliminate or reduce many diseases in densely populated areas, and contribute to the development of urban life. Additionally, many Mediterranean regions, such as Provence in the south of France, or Castile in Spain, Florence in Italy, earn money by producing soap and cosmetics due to the rich resources of natural olive oil they have [6]. Olive oil is rich in palmitic, stearic and unsaturated oleic, linoleic, and unsaturated linolenic acids, which leads to the production of quality soaps. Sulfate oils were the first synthetic surfactants to be prepared after soap in 1834 by Runge via mixing olive oil and sulfuric acid. In 1875 sulfate castor oil, also known as Turkey red, was prepared as a dye additive. Additives were used in the textile industry [7]. Sulfate oil is not a pure surfactant, but a mixture of sulfate esters, water and fatty acid surfactants. In 1935, Colgate-Palmolive developed the first soap-free shampoo using mono surfactants and sulfated diglycerides [7]. Despite the long history of amphibians, these compounds are still of interest to scientists, who are looking to design new types of amphiphiles for use in a variety of fields.

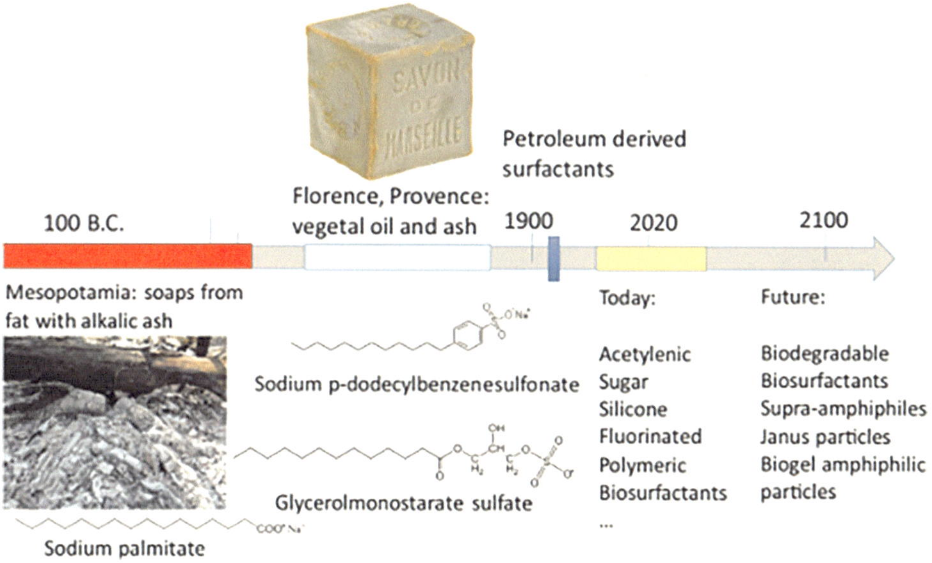

Figure 3.
Time axis for the evolution of surfactants toward the present and future. Sources: Ash photo: (Pile of ashes after the fire went out. Ash texture. Burned out ashes grunge texture. Australia Shutterstock stock photo ID: 1611309511 (https://www.shutterstock.com/image-photo/pileashes-after-fire-went-out-1611309511)). Soap photo Soap of Marseille Royalty-free stock photo ID: 60864694 (https://www.shutterstock.com/image-photo/soap-marseille-isolated-60864694).

3. Types of amphiphiles

As mentioned earlier, amphiphilic molecule is a general term for a compound that contains two parts with different covalent bonds and affinity for polar and non-polar solvents. The polar part has a high affinity for polar solvents (such as water) and the non-polar part has a high affinity for non-polar solvents such as hydrocarbons, ethers and esters. Surfactants, simple fatty acids, polymeric amphiphiles, and some lipid molecules, containing hydrophilic and hydrophobic components, are typical examples of amphiphilic molecules. Now a dayes, due to the use of amphiphilic compounds in various industrial and medical fields, scientists are looking to design and synthesize new types of these compounds. The hybrid of conventional building blocks with biomolecular and mineral components has led to new structures of these compounds with unique self-assembly properties. For example, the combination of an alkyl chain and a short peptide sequence results in the synthesis of amphiphile peptides that can aggregate spontaneously into various one-dimensional structures [8]. Other examples include giant amphiphiles from a single-protein hybrid [9]. Polygonal oligomeric silescoyoxanes [10] or polyoxometetates [11], which spontaneously accumulate in new structures such as toroids and other complex architectures, show interesting features that are significantly different from ordinary amphiphiles. Interestingly, these new designs have led to the synthesis of giant amphiphiles (supraamphiles). Unlike conventional amphiphiles based on covalent bonds, new amphiphiles are designed based on non-covalent interactions or dynamic covalent bonds [12–15]. In supraamphile, functional parts can be connected by non-covalent bonds. In this way, these compounds can be synthesized without the need for boring chemical syntheses. Building blocks for superaphiphiles can be small molecules or polymers. The development of supraamphiphiles not only enriches the common amphiphilic family, but also provides a new type of building block for complex self-assemblies, including hierarchical self-assemblies and functional nanostructures. Since surfactants are the most obvious and widely used group of amphiphiles, after a brief review of the types of amphiphiles, we will try to explore this group of amphiphiles in more detail. Amphiphilic molecules can be divided into two major categories based on the type of bond that connects all parts of the molecule as follows.

3.1 Conventional amphiphiles

In this type of molecule, all the parts are connected only by covalent bonds.

3.1.1 Bioamphiphiles

Many biological molecules, such as proteins, phospholipids, cholesterol, glycolipids, and bile acids, are also made up of two parts, hydrophilic and hydrophobic, and have a dual amphibian property called bioamphophiles.

3.1.1.1 Amphiphilic proteins

Amphiphilic proteins are composed of sequences of polar and non-polar parts of amino acids. For example, a protein may consist of hydrophilic moieties of polar amino acids (such as Asp-Ser, Tyr-Glu) and hydrophobic moieties of nonpolar amino acids (such as Gly-Pro, Ile-Pro-Met). This is the structure of membrane proteins in biological membranes. Their hydrophobic nature causes them to place themselves in the hydrophobic and non-polar regions of a biological membrane and expose their hydrophilic part to the polar environment. These hydrophilic parts of the protein lead to their interaction with polar molecules. Most of these amphiphilic proteins have these seemingly opposite interactions due to their amphiphilic helix structure. An amphiphilic helix is a protein helix that has two opposite sides. The face along the long axis of the spiral is hydrophilic, while the opposite face is hydrophobic. Thus, it can separate the hydrophobic and hydrophilic parts of a protein, resulting in a protein–protein interaction. Amphiphilic helices are a common structural feature in proteins. Ion channel membrane proteins, lung surfactant proteins, and apolipoproteins are examples of proteins with this structure [16].

3.1.1.2 Phospholipids

Phospholipids are also bio-amphiphilic molecules which are lipids with a glycerol group attached to two fatty acids and a phosphate group. Glycerol forms a phospholipid hydrophilic head by binding to a negatively charged phosphate group. The phosphate group may bind more to hydrogen, choline, serine, ethanolamine, or inositol, thus converting to phosphatidic acid, phosphatidylcholine, phosphatidylserine, phosphatidylethanolamine, and phosphatidylinositol phosphatidyl, respectively. Two long chains of fatty acid form the hydrophobic tail of phospholipids, which is lipophilic. The amphiphilic nature of phospholipids has made them an essential component of biological membranes. For example, the plasma membrane consists mainly of two phospholipid layers. Phospholipids can interact with different molecules depending on their polarity. Phospholipid heads readily interact with water and other polar molecules. Conversely, phospholipid tails avoid water and other polar interactions. Thus, phospholipids in water accumulate by directing their tails toward each other while exposing their heads to the aquatic environment. In fact, it is the amphiphilic nature of phospholipids that leads to the formation of the bilayer structure of the plasma membrane. Phospholipid tails are oriented in such a way that their tails are placed inside the plasma membrane while the head of phospholipids is placed outwards.

3.1.1.3 Cholesterol

Cholesterol, which is composed of a hydrophilic hydroxyl group (-OH) and a bulky hydrophobic steroid and hydrocarbon chain, is also a bioamphiphile. Cholesterol is found in the plasma membranes of animals. Its hydrophilic part interacts with the aqueous medium and with the polar heads of phospholipids, and its hydrophobic part, in turn, is located next to the hydrophobic tail of phospholipids and the chain of non-polar fatty acids of other lipids in the membrane.

3.1.1.4 Glycolipids

Glycolipids, which are composed of hydrophilic sugar groups attached by covalent bond to the hydrophobic lipid tail, are considered amphiphiles. They are also present in plasma membranes like cholesterol. Their carbohydrate portion extends to the outer surface of the cell while the lipid portion lies in the lipid bilayer. The residues of the sugar group, which are located on the outer surface of the cell, allow carbohydrate-carbohydrate interaction.

3.1.1.5 Bile acids

Bile acids are steroid-structured amphiphiles consisting of four rings and a side chain ending in carboxylic acid and hydroxyl groups. Bile acid salts, like surfactants, can form micelles by accumulating around lipid droplets. They can also emulsify lipids and prevent fat droplets from accumulating into larger fat particles.

3.1.1.6 Saponins

Saponins, which are abundant in plants, are made up amphiphilic compounds composed of a glycoside, which forms their hydrophilic moiety, and a triterpene, or steroid derivative, which is their hydrophobic moiety. These compounds are bitter and toxic and reduce the good taste of plants.

3.1.2 Polymer amphiphiles

Polymer amphiphiles are copolymers that consist of a block as a hydrophobic tail and a block as a hydrophilic head. While most copolymer amphiphiles are composed of two blocks, some may consist of three blocks and other geometric shapes and exhibit amphiphilic behavior. There are many polymer amphiphiles, for example, Pluronics, also known as poloxamer, is a synthetic block copolymer composed of hydrophilic poly (ethylene oxide) (PEO) and hydrophobic poly (propylene oxide) (PPO), which are located in a triple structure A-B-A, and form PEO-PPO-PEO.

3.1.3 Surfactants

According to **Figure 4**, surfactants are usually organic compounds that are amphiphilic. These compounds are synthetic or biological (biosurfactant) or may be extracted from natural materials such as some plants (glucosides or natural surfactants). Careful studies show that surfactants have unique applications in nanotechnology, such as their use in the synthesis, modification of properties, stabilization, self-aggregation, drug deliveries, and so on. Therefore, it can be said that they play a key role in empowering future science and technology. Surfactants are promising new areas of research in various industries of the future, as they can be used in the preparation of important materials that form the basis of various industries, such as carbon nanotubes or graphene. What makes surfactants attractive is that, unlike other molecules that remain there after being dissolved in water, they will always look for an adsorption interface after dissolution, whether solid–liquid, liquid–liquid, or liquid–gas. They can be said to be a kind of "smart" molecule. This behavior leads to their ability to assemble into extraordinary structures called self-aggregation. For this reason, in this chapter, among the different types of amphiphiles, surfactants will be explored in detail as part 3.

3.2 Advanced amphiphiles

These compounds are also called supermolecular amphiphiles (or superamphiphyls). They are a new bridge between colloidal science and supramolecular chemistry and provide a context in which we can make full use of our imagination. In these amphiphiles, all segments are connected by non-covalent such as

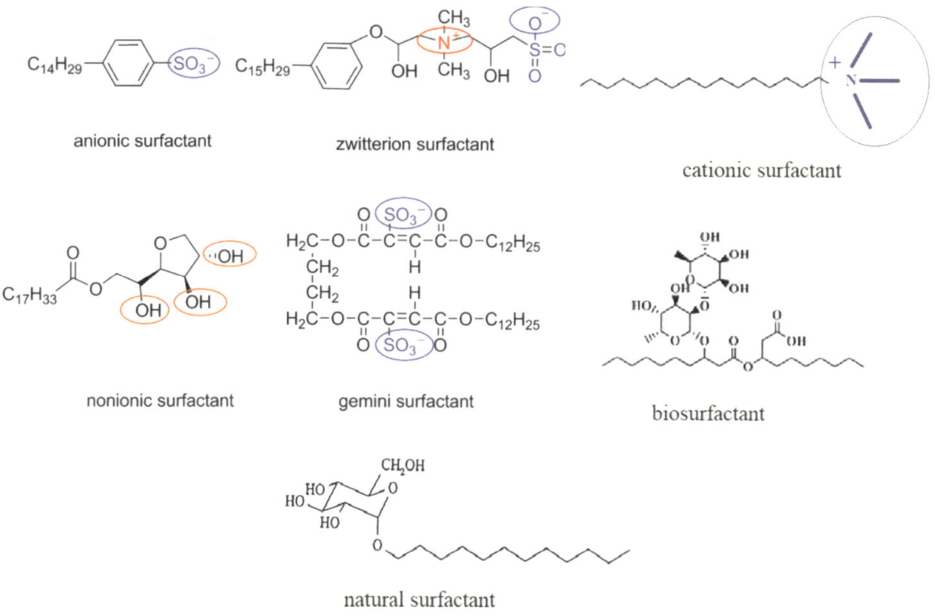

Figure 4.
The chemical structure of some of surfactants.

electrostatic interaction, π-π stacking, charge-transfer interaction, hydrogen bonding, and host–guest interaction [14, 15] or dynamic covalent bonds such as imine and disulfide bonds, which are similar to noncovalent interactions under certain condition [17, 18]. In other words, these types of amphiphiles, unlike conventional amphiphiles, have hydrophilic and hydrophobic parts that are connected by physical bonding; consequently, these compounds and their aggregations are smart to external stimuli such as gas, light, temperature, etc.

3.2.1 Hybrid amphiphiles

Hybrid amphiphiles are compounds whose hydrophilic and hydrophobic moieties can be organic or inorganic compounds that are joined by non-covalent or dynamic covalent bonds. These compounds can exhibit additional functional properties such as special magnetic or catalytic properties or new properties that none of the components alone can have [19].

3.2.2 Janus amphiphiles

The word "Janus" in Greece is a word that is the opposite of itself. Thus, Janus amphiphiles, which are usually much larger than conventional surfactants, have both hydrophilic and hydrophobic components on the same (hard) particle in their structure. In recent years, Janus amphiphile particles have attracted a lot of attention due to their asymmetric and multifunctional structure [20–24]. In contrast to particles with uniform surface wettability, the surface of Janus amphiphile particles is divided into two parts with different wettability. Since Janus amphiphile particles have a unique structure, they exhibit special behaviors.

4. Principal types of surfactants

Surfactants are classified in different ways based on the chemical structure of the head group, the chemical structure of the tail group, the electrical charge of the head group, the type of application in industry, and the raw material from which they are derived.

On the one hand, the structure of the components of surfactants affects their application in various fields, thus its study is of particular importance. On the other hand, this book focuses on the self-assembled structures of materials and for the accumulation of surfactants, the chemical structure and charge of the head group and the chemical structure of the tail group are important, hence in this section we try to classify them from this perspective.

4.1 The electrical charge of the head groups

4.1.1 Anionic

Anionic surfactants are surfactants that have a negatively charged head group. In fact, any negatively charged organic compound with a positively charged ion (-counter-ion) introduces an anionic surfactant.

4.1.2 Cationic

Cationic surfactants are surfactants whose head group has a positive charge. This charge may be permanent or only within a certain pH range. According to **Figure 5**, the negative charge of anionic surfactants is mostly on the oxygen atom and is available, while the positive charge of cationic surfactants is on the nitrogen group surrounded by alkyl groups. Consequently, the distances from the micellar core-water interface to the charges on anionic (SDS) and cationic (CTAB) surfactants, $d_{ch,SDS}$ and $d_{ch,CTAB}$, respectively, need not be the same.

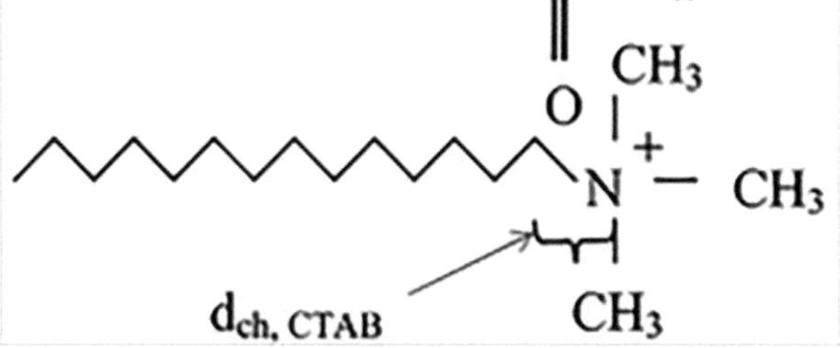

Figure 5.
The distances from the micellar core-water interface to the charges on anionic (SDS) and cationic (CTAB) surfactants, dch, SDS and dch, CTAB.

A new model has been proposed for estimating the surface potential and electrostatic free energy (g_{elec}) using the capacitive model due to the difference in the location of the charge of anionic and cationic surfactants. Compared to the work of others, it is applicable to accurately study the changes in electrostatic energy and free transfer energy in transition of the accumulation of surfactants to each other [25].

4.1.3 Zwitterionic

Zwitterionic or amphoteric surfactants are surfactants whose functional group has both positive and negative charges. Like cationic surfactants, these charges can be permanent or pH dependent.

4.1.4 Nonionic

Nonionic surfactants are surfactants with polar head groups without electric charge. They usually have a functional group that can be slightly deprotonated. Therefore, despite its good solubility in polar solvents such as water, it does not work well as a Brønsted acid. In general, nonionic surfactants, unlike ionic surfactants, do not have good solubility and cannot change the pH of the solution.

4.2 The chemical structure of the head groups

4.2.1 Carboxylic acid salt, $RCOO^- M^+$

Carboxylates or carboxylic acid salts are surfactants in which the head group (hydrophilic group) is a carboxylate group, COO^-. **Figure 6** shows the structure of some of these surfactants.

4.2.2 Sulfonic acid salt, $RC_6H_4SO_3^- M^+$

The second most important class of anionic surfactants are sulfonic acid-based surfactants. **Figure** 7 shows some of the most important compounds in this class. One

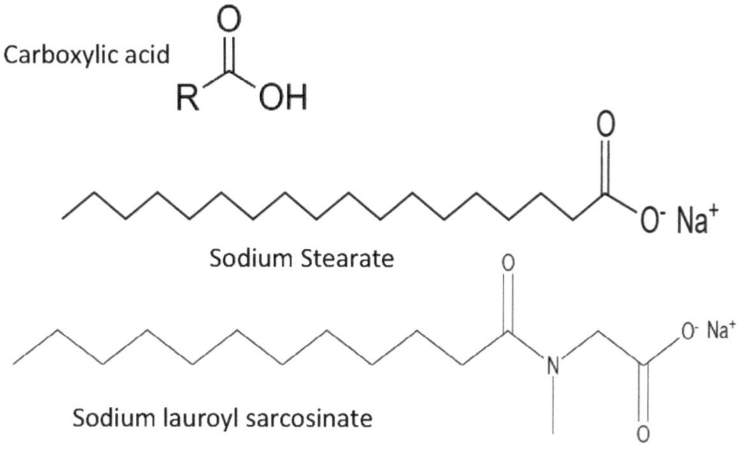

Figure 6.
The structure of some of carboxylic acid salt surfactants.

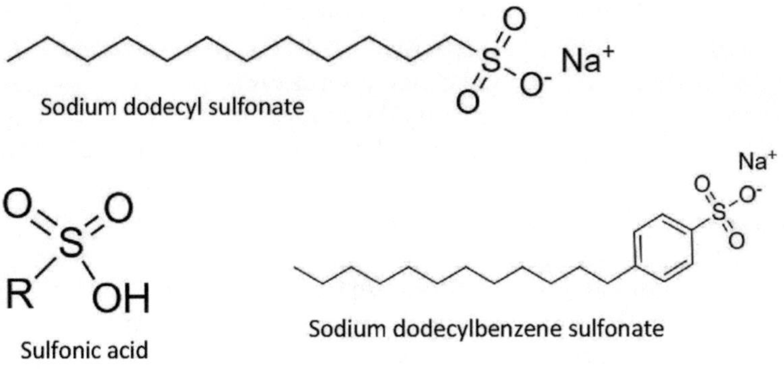

Figure 7.
The structure of some of sulfonic acid salt surfactants.

of the most important surfactants in this category is sodium dodecyl sulfate, which is the sodium salt of dodecyl sulfonic acid. This surfactant has many industrial and medical applications. In addition to the surfactant made from aliphatic sulfonic acid, sodium dodecyl benzene sulfonate, one of the most common anionic surfactants, is made from aromatic sulfonic acid.

4.2.3 Sulfuric acid ester salt (sulfate salt), $ROSO_3^-M^+$

Sulfate salts consisting of a predominantly linear alkyl chain having a terminal sulfate ester anion neutralized with an opposite ion. The most common structural feature of these surfactants is the presence of a predominantly linear aliphatic hydrocarbon chain with a neutral polar sulfate with an opposite ion (e.g. Na^+, K^+, NH_4^+, or an alkanolamine cation; **Figure 8**). The hydrophobic hydrocarbon chain (with lengths between C8 and C18) and the polar sulfate group provide surfactant properties in these compounds and allow the commercial use of these materials as anionic surfactants.

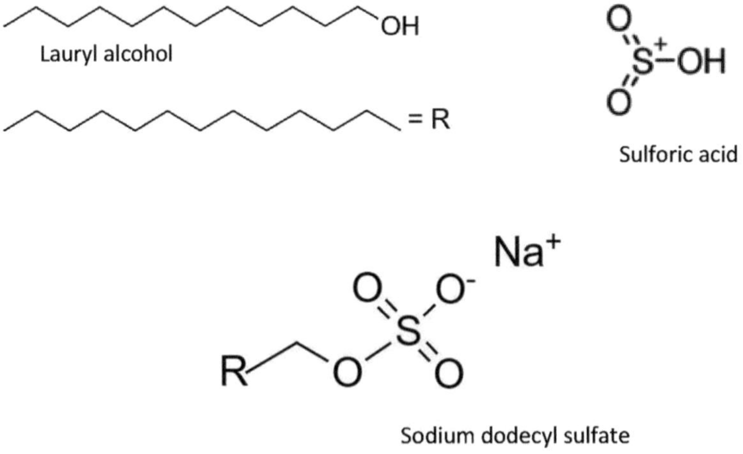

Figure 8.
The structure of sulfuric acid ester salt surfactants.

4.2.4 Phosphoric and polyphosphoric acid ester, $R(OC_2H_4)_xOP(O)(O^-M^+)_2$ and $(R(OC_2H_4)_xO)_2P(O)O^-M^+$

Anionic surfactant-phosphate ester-contains phosphate fraction as a hydrophilic head. Phosphate ester surfactants are produced by the reaction of alcohols with a phosphoric acid derivative. The property of phosphate ester depends in part on the structure of the primary alcohol used in its preparation. From the reaction of phosphoric acid with alcohol, polyphosphate ester surfactants can be prepared. **Figure 9** shows the structure of anionic surfactants of phosphate ester and polyphosphate ester.

4.2.5 $RNH_3^+Cl^-$ (salt of a long-chain amine), $RN(CH_3)_3^+\ Cl^-$ (quaternary ammonium chloride)

Cationic surfactants have almost exclusively nitrogen-containing functional groups such as amines, quaternary ammonium, and almost imidazolium, and more recently pyridinium functional groups (**Figure 10**). Ammonium surfactants are active and at low pH values when the amine group is protonated, are cationic and hydrophilic. Thus, the amine group is a potential factor in surfactants for making pH-responsive amphiphiles.

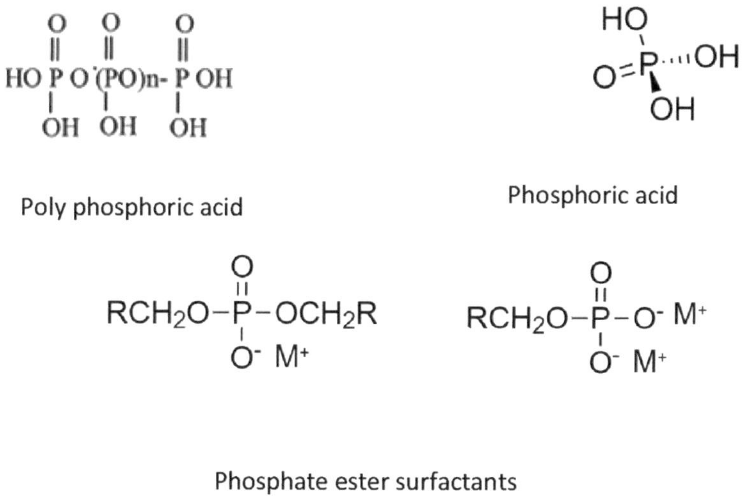

Figure 9.
The structure of phosphoric and polyphosphoric acid ester surfactants.

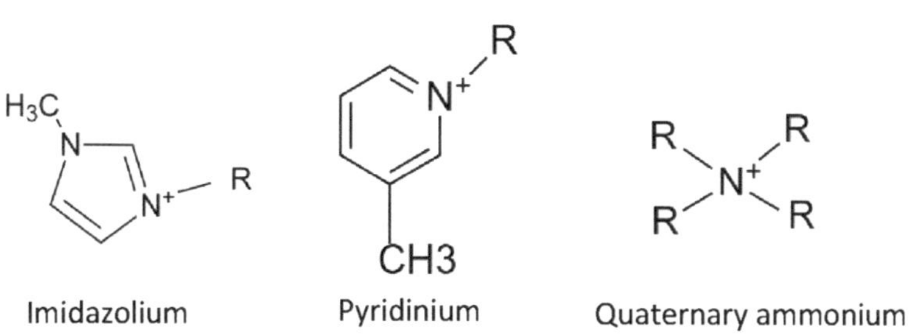

Figure 10.
The structure of some of cationic surfactants.

4.3 The chemical structure of the tail groups

The chemical structure of the tail group of surfactants is divided into perfluoroalkyl, alkyl, alkyne, silicon, saturated, unsaturated, etc.

5. Surfactants self-assembly in solution

In the hydrophobic phenomenon, the hydrophobic tail of a surfactant molecule is not soluble in water due to the presence of hydrogen bonds between water molecules and the lack of strong bonds between surfactant molecules and water molecules. According to **Figure 11**, when the surfactant is placed at the air/water interface, its hydrophilic head group is in the water (liquid) phase, and its hydrophobic tail is in the air, which is a more suitable environment for it. Since it takes less energy to bring a surfactant molecule to the interface than a water molecule, adsorption is a spontaneous process. In other words, the surfactant molecule overcomes the hydrogen bond between

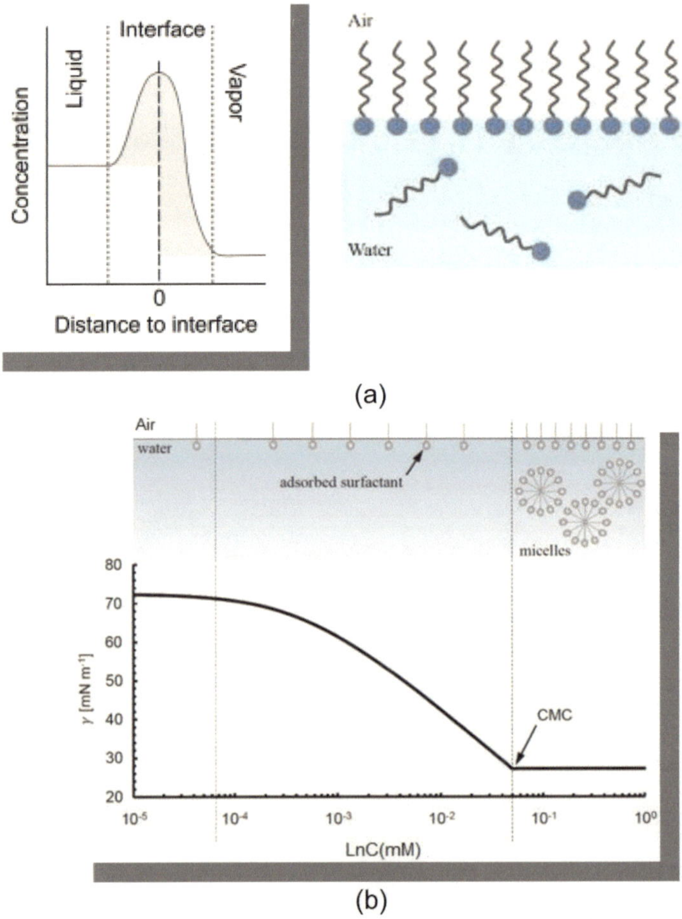

Figure 11.
The schematic of (a) surfactant adsorption at the air-water interface and (b) surfactant aggregation in bulk solution.

the water molecules at the water surface by adsorption at the interface, thereby reducing surface tension. For each surfactant, depending on the structure and charge of the head group and the structure of its tail group and according to environmental conditions at a certain concentration after saturation of the interface, the tail of surfactant molecules assembles in a water bulk and forms aggregation. This specific concentration is called the critical micelle concentration (CMC). Since the sphere is the most stable geometric shape, the first micelle formed will be spherical. With the tail accumulation of surfactants together, the entropy of the system decreases thermodynamically. This accumulation, which occurs mainly due to the hydrophobicity of the tail group of surfactants, minimizes their undesirable contact with water molecules because it is more suitable for the tailings of surfactants to be placed together in bulk. As a result, the combination of these factors will lead to the stability of the system and the formation of aggregation. In other words, although the surfactant phase accumulates more regularly (lower entropy), water molecules that had an unfavorable orientation around the surfactant sequences are now released, leading to a net increase in the net entropy of the universe. Therefore, the process of micelle formation is a spontaneous process. At concentrations below CMC, the adsorption of the surfactant molecule at the interface between air and water depends on the surfactant concentration, and increasing the concentration leads to an increase in adsorption and thus a decrease in surface tension. However, after reaching the CMC, the additional surfactant molecules that are added to the bulk phase simply form the micelles, because this path is energy efficient and the surface tension of the liquid phase will remain constant after the CMC point and does not alter with the concentration changes.

5.1 Aggregation and packing parameter

As mentioned earlier, the CMC point is the first concentration at which it is thermodynamically preferred that surfactants accumulate in the form of spherical micelles. This is a phenomenon known as self-aggregation, and almost all amphiphiles spontaneously form regular structures that are thermodynamically stable. In these self-aggregating structures, the aggregation number is an important parameter. In fact, it shows the average number of surfactants in a micelle. On the one hand, experimental studies prove that the radius of the micelles cannot be longer than the length of the alkyl surfactant chain. On the other hand, these studies show that the diameter of the micelle is very close to the length of its alkyl chain, so the inside of the micelle is very compact and water or any other polar liquid cannot be placed in it.

The aggregation number can be obtained in two ways:

1. From the ratio of the volume of micelle to the volume of a surfactant molecule:

$$N = \frac{v_{mic}}{v_{sur}} = \frac{4\pi R_{mic}^3}{3v_{sur}} \tag{1}$$

2. From the ratio of micelle area to the area of the surfactant:

$$N = \frac{a_{mic}}{a_{sur}} = \frac{4\pi R_{mic}^2}{a_{sur}} \tag{2}$$

The above two terms are equal thus we have:

$$\frac{v_{sur}}{R_{mic}a_{sur}} = \frac{1}{3} \tag{3}$$

This parameter, which is $\frac{1}{3}$ for spherical aggregates, is called the packing parameter. As the concentration of surfactants increases, the size of the micelles first increases, but since the sphere can be somewhat stable, it either breaks into smaller spheres or translates into rod, cylindrical, bilayer, or cyclic structures. The tendency of surfactants to form different phases and structures in solution is controlled by the interactions between the surfactant head groups, the tail and solvent groups, and the geometric factors. The packing parameter mentioned above can be used to predict the shape of assemblies before preparing them experimentally. The surfactant packing parameter was introduced by Israelachvili et al. [1] and provides a criterion for predicting the shape that aggregates of a given surfactant will adopt in aqueous solutions at a concentration well above the critical aggregate concentration (CAC). The packing parameter, P, is

$$P = \frac{v}{a_0 l_c} \tag{4}$$

This parameter indicates that, in the structure of a typical surfactant, the alkyl chain has the largest segment in the volume of the molecule and the dimensions of the micelle, and the hydrophilic head group exclusively contributes to the interface of the micelle and water. Therefore, the above equation can be interpreted as the volume of alkyl chains divided by the area of the hydrophilic group at the common level multiplied by the length of the alkyl chain.

According to Tanford equations [26], the hydrophobic chain volume of the micelle, v, and the critical chain length, lC, can be obtained from.

$$v = (27.4 + 26.9n_c)\text{Å}^3 \tag{5}$$

and

$$l_c = (1.5 + 1.265n_c)\text{Å} \tag{6}$$

where n_C is the number of carbon atoms in the hydrophobic chain of the surfactant. It has been proposed [27] that the surface area per headgroup, a_0, is the most important factor controlling aggregate size. According to **Figure 12**, the packing parameter determines whether micelles are spherical (P < 1/3), nonspherical (1/3 < P < 1/2), vesicles or bilayers (1/2 < P < 1), or [28] "Inverted" structures (P > 1).

In fact, the packing parameter is responsible for the curvature of the self-assembled structure based on the surfactant geometry. Studies show that for charged ionic groups, a_0 is not the area occupied by the hydrophilic group, but the area of interaction of the two electrical layers determined by the length of Debye. The volume of the hydrophobic micelle core is mainly affected by the branching and length of the alkyl chain. The main weakness of the compaction parameter, which limits its practical use, is that first the type of self-assembly structures produced by a surfactant must be determined, then the compaction parameter must be calculated. For example, when this parameter is $\frac{1}{3} < P < \frac{1}{2}$, the spontaneous curvature of the self-cumulative structure as shown in **Figure 12** should be spherical with a slight curvature. While this

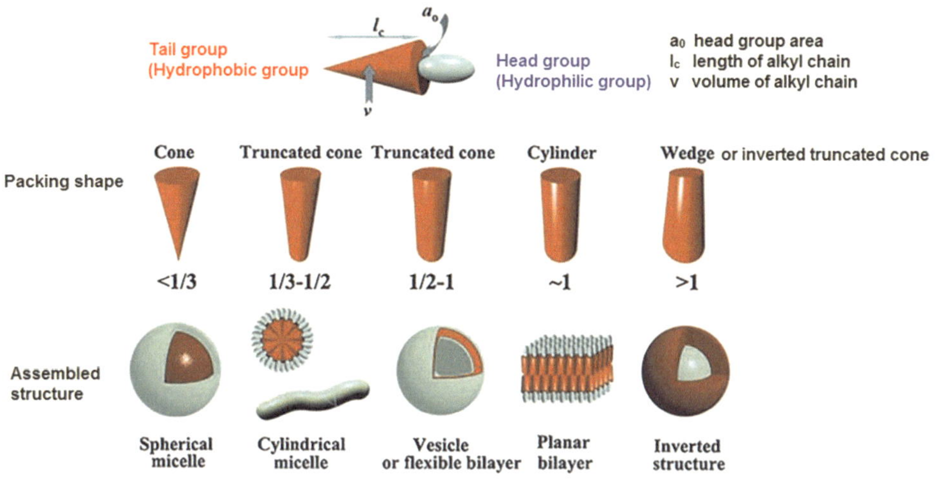

Figure 12.
Graphic representation of individual surfactant geometry and the corresponding aggregate structure [29], with permission from Elsevier.

structure does not form because there should be no water inside the hydrophobic micelle core, there is an empty space under extreme hydrostatic pressure that causes the spherical structure to become a cylindrical structure for further contact with the alkyl chains, which causes stress. Closure is at the end of cylindrical, rod-shaped or worm-shaped micelles (**Figure 12**). Closure stress in this type of micelles is the reason that rod-like and cream-like micelles grow with increasing surfactant concentration and eventually become surfactant mesophase [28]. The formed micelles can be observed through electron microscopy studies [30, 31]. The geometry of the surfactant can be adjusted not only by chemical design, but also by adjusting the area of the hydrophilic head group, which is related to the length of Debye. Debye length is affected by counter-ion valence, ionic group load, salt concentration [32], and so on. In this way, the effect of surfactant geometry on the shape of the aggregates can be adjusted by adding salt [33].

Eventually, the two most important parameters for identifying a surfactant and preparing its solution are the Kraft point and the cloud point. The Kraft point is the temperature at which the value of CMC is the same as the solubility of the surfactant. The Kraft temperature for a given hydrophilic group increases with increasing carbon content in the hydrophobic group and decreases with branching. For a given hydrophobic group, the Kraft temperature decreases with increasing number of hydrophilic groups and for ionic surfactants decreases with increasing degree of counter-ion hydration [34]. To prepare a low temperature surfactant solution, it is critical to select surfactants with a low Kraft temperature. Cloud point is the temperature at which non-ionic surfactants in water are insoluble, resulting in the solution becoming a cloudy suspension that scatters light.

5.2 Inter-aggregation (long distance) and intra-aggregation (short distance) interactions

In this section, we will try to examine the interrelationships between the interactions that are responsible for the self-aggregation of surfactant molecules and amphiphiles and the interactions that occur between aggregations as they approach each other.

Essentially, there are four types of interparticle interactions: hard-sphere, soft (electrostatic), van der Waals, and steric. The hard-sphere interactions are repulsive and only significant when the particles are close to each other at distances of slightly less than twice the radius of the hard sphere [35].

5.2.1 DLVO theory: soft (electrostatic) and van der Waals interactions

The interactions between soft particles (electrostatic) and van der Waals (vdW) are considered in a theory independently proposed by Derjaguin and Landau [36] and Verwey and Overbeek [37], hence the so-called DLVO theory. According to this theory, in order for particles to disperse, a balance must be struck between attractive and repulsive energies. This theory states that the stability in the solution of surfactant (colloidal solution) is determined by the sum of the electrostatic repulsion forces associated with the double layer around the aggregates formed by these materials and the attraction and vdW forces between them. This theory suggests that a potential barrier to repulsive force prevents two particles from coming close to each other and adhering to each other. Therefore, if the repulsive particles are high enough, the dispersion will resist coagulation and the colloidal system will be stable.

Repulsive interactions are believed to be either due to double electrical layers with the same charge surrounding the particles or due to particle-solvent interaction. In other words, attractive interactions are mainly due to van der Waals forces between particles. For particle dispersion, repulsive interactions must be large enough to overcome attractive interactions. To accumulate particles, it must be reversed.

5.2.2 Steric interaction

One of the effective interactions in the formation of self-assembly structures is the interaction caused by steric hindrance [35], which is also known as non-DLVO interaction. This interaction creates two effects: (1) mixing effect and (2) entropic effect. The mixing effect is due to solvent-tail interactions and the high concentration of tails in the tail-solvent contact area. This effect becomes noticeable when the interaction of the tail of the surfactant with the solvent is stronger than the hydrophobic interaction between the tails, in this case the tendency of the particles to accumulate will increase and then the free energy of the system will decrease. In the other hand, if the tail-tail interaction is more than the solvent-tail interaction, then the hydrophobic interaction between the tails leads to the formation of aggregation. The entropic effect is due to the limited motion of tails that expand in the liquid phase when adjacent particles approach each other.

Both effects lead to an increase in the accumulation in the liquid phase by increasing the number of tails and increasing their length. However, the length of the tails should be optimized, because the possibility of aggregation increases with the increase in the length of the surfactant tail and the increase in the hydrophobicity phenomenon.

In cases where the liquid phase is a liquid other than water, the steric hindrance has the best effect on the aggregation when the head group has limited solubility in the liquid phase, as a result, despite the steric hindrance of the tails, monomers prefer to accumulate to reduce the free energy of the system. In general, it can be concluded that since the formation and deformation of self-assembled structures of surfactants largely depends on electrostatic, vdW, hydrophobic and steric interactions [38, 39] and the balance between them; consequently, structural parameters such as head group, chain length and counter ion on the surface charge density of micelles and the interaction between them and as a result have an effect on the size and shape of the aggregates. Also, environmental effects including ionic strength, temperature, electrolyte conditions such as salts and ionic liquids also have an effect on the interactions and as a result the size and shape of the aggregates.

Generally, it is possible to control the CMC parameter and the packing parameter by properly designing the head and tail structure of surfactants in such a way that in addition to reducing the CMC, different forms of aggregations can be produced based on their application. For example, atoms such as nitrogen or oxygen that can form intramolecular hydrogen bonds can be used in the surfactant head group structure. Investigations show that the formation of intramolecular hydrogen bonds leads to a decrease in CMC and an increase in aggregation number [40].

Since ions with the opposite charge are attached to the micelle surface, the nature of these ions affects the properties of micelles. For example, the size of micelles formed with a cationic or anionic surfactant increases with increasing counter ion size. Micelle size is not only a function of the size and electronegativity of the counter-ion, but also depends on the size of the hydration layer surrounding the counter-ion. So that weakly hydrated ions (smaller and highly electronegative) that are adsorbed closer to the micelle surface can neutralize the surfactant charge more effectively and lead to the formation of smaller micelles.

6. Complex self-assemblies beyond the conventional self-assemblies

The chirality of a compound, which is its lack of mirror symmetry, is a fundamental concept that affects the assembly, creation, and biological, chemical, mechanical, and optical properties of materials, and is intricately involved in the structure of life. Helical self-assembled structures with controllable twists have recently attracted considerable attention because they mimicks biological helices and can be used to design new chiral materials. However, coordination polymers (CPs) with helical morphology have rarely been discovered so far. In particular, chirality inversion via external actuation has not been achieved in helical CPs. Based on this, Zhang et al. have been able to synthesize coordination polymers (CPs), superhelices, which consist of metal ion nodes and organic linkers. In this study, they succeeded in providing a new way to fabricate helical CP nanostructures with desirable chirality, which have potential applications in chiral catalysis, chiral separation, and chiroptics [41]. In another work done by Diao et al., they were able to prepare hierarchical helical structures through a multi-step aggregation pathway [42]. Using experimental techniques, they were able to show that increasing the concentration of the solution leads to the production of multi-scale helical aggregation of twisted polymers. In other words, polymers progress from the nano scale to the micron scale. This study could pave the way for the field of chiral optoelectronics. Because the emergence of hierarchical chirality in non-chiral conjugated polymers can profoundly affect the way these polymers interact with light and the transmission of signals resulting from biomolecular interactions and cause properties that were not even imagined before. These new and advanced applications of complex self-assembly structures prompted us to further study these compounds.

In the previous sections, the phenomenon of self-assembly and the factors affecting it were explored in detail. This phenomenon forms the basis of many natural and biological processes such as protein folding and cell membrane formation. These processes are controlled by complex intermolecular and intramolecular interactions such as metal coordination, hydrophobic effect, hydrogen bonds, electrostatic interactions, π–π interactions and van der Waals forces. While the self-assembly of amphiphiles into conventional structures such as micelles or vesicles has attracted much attention for many decades due to their wide application in materials science, medicine, and gene delivery, with recent advances in nanoscience, the way to the emergence of supramolecular self-assembly structures has opened up the development of complex and hierarchical structures. In the other words, amphiphilic molecules, in addition to micelles and vesicles, according to **Figure 13** can be hierarchically assembled in the form of more regular nanostructures such as fibers, helices, superhelices, ribbons and tubes, which are of interest to supramolecular chemists. In these new nanostructures, despite knowing the basic modes of supramolecular interaction, it is difficult to predict the shape of the self-assembly system. So far, studies have shown that some self-assembled structures are formed by creating a balance between hydrophilic and hydrophobic components, while recent studies have shown that chirality may play an important role in controlling the shape of self-assembled nanostructures. Hence, the phenomenon of chirality can play an important role in the complexity of these structures and functions at the molecular level. Also, these complexities have

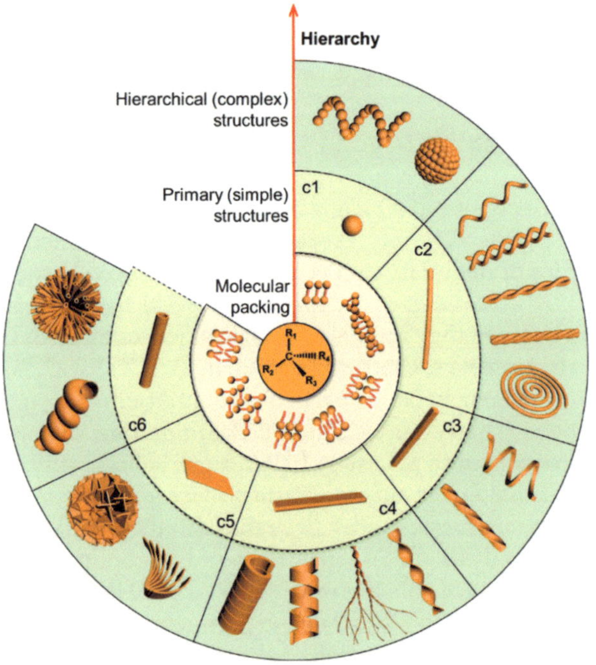

Figure 13.
The diversity of chiral nanostructures constructed through hierarchical self-assembly. Starting from the molecular level, various primary structures can be formed through plausible packing modes. Hierarchical self-assembly of higher order or complex structures can be attained by using these primary nanostructures as building blocks. In the clockwise direction, (c1) zero-dimensional nanospheres can form helical sphere chains and microspheres; (c2) one-dimensional nanofibers can further form single-helical fibers, double-helical fibers, helical bundles with two fibers and multiple fibers, and chiral spirals; (c3) one-dimensional nanorods can assemble into helical and twisted nanorods; (c4) two-dimensional nanobelts can form twisted ribbons, dendritic twists, helical ribbons, and chiral tubules; (c5) two-dimensional nanosheets can assemble into helically arranged petals and microspheres; (c6) three-dimensional nanotubes can form microtubule flowers and helical tubes [43, 44]. With permission from Royal Society of Chemistry.

been observed in right-handed alpha-helical structures of proteins or double-helix DNA. Research has shown that these helical structures can be single, double, triple or quadruple strands and exist as linear or circular sets [45].

Some research has shown that chirality can play an important role in controlling the shape of self-assemblies such as tubes and the morphology of helical ribbons. In order to design new supramolecular self-assembly structures for using in nanotechnology, in addition to understanding the form of chiral self-assembly, it is necessary to control the size and shape of the aggregates by adjusting the chemical composition, interactions, and conditions under which self-assembly occurs. In this regard, the important step is to know how the three parameters of the elastic force, chirality and the direction of membrane inclination control the structure and shape of the resulting aggregation. Previously, presented models show that aggregates of cylindrical tubes or spiral bands are formed due to chiral interactions, along with molecular tilt [46]. Since this model only predicts the formation of ideal structures, as a result, other models have been presented, including the theory that has been proposed for the effect of chirality on the shape of nanostructures resulting from complex supramolecular self-assembly [47–49]. According to these theories, the existence of some chiral restrictions in the systems leads to the non-parallel arrangement of macromolecules during the formation of self-aggregates. In other words, the macromolecules are forced to accumulate at a non-zero angle near to their closest neighbor, as a result, they will prefer a specific orientation in the double layer structure, which causes double layer deviation and as a result, the formation of a hollow cylinder will be followed by a fibrous morphology.

Nowadays, metamaterials with better mechanical, optical and even electromagnetic properties can be produced by manipulating the morphological chirality. Also, studies have shown that chirality has a significant effect on physiological properties and medicinal effects, so it can be said that chirality has been proposed as one of the most important research from a biological and chemical point of view. However, many features in nature, such as tunable and hierarchical chirality, which can provide electromagnetic control of light polarization and enhancement of mechanical properties in man-made structures, are still an important challenge. In this section, an attempt has been made to examine the concept of chirality and how it changes during the formation of aggregate structures and the parameters affecting it, so that the way to solve the existing challenges may be smoothed. Parameters such as the intensity of interaction between components in the aggregate structure, their electronic properties, steric effects, geometric preferences, stoichiometry, as well as external factors such as polarity, temperature, and pH play a central role in how chirality affects the shape of aggregates [50]. In 2020, Chen et al. have provided inspiration from origami techniques [51]. They have been able to propose egg-box-based chiral units for the construction of homogeneous and heterogeneous chiral structures and show a theoretical approach to adjust the chirality of these structures by adjusting their geometrical parameters and to achieve chirality change through the branching mechanism. Gradually, they placed a helical bond between the chiral units and were able to design hierarchical structures with chirality that is transferred from the structural elements to the morphological level. Their proposed method can lead to the development of artificial metamaterials with chiral properties, further these metamaterials may be used in engineering applications, including switchable electromagnetic metamaterials, morphing structures, and bionic robots. Put simply, chirality indicates the asymmetric property of an object that has at least one asymmetric center in its structure, in which case the object cannot conform to its mirror image. On the scale of macrochirality, it can usually be morphologically represented by a helix, Create a structure that exists in most biological and synthetic materials. The presence of chirality in materials leads to unusual mechanical, optical and electromagnetic properties and functions [52–55], such as negative refraction, asymmetric electromagnetic waves transmission and superior elasticity.

Investigations show that chirality translation, in other words, the conversion of right-handed and left-handed chiralities into each other is an important phenomenon. On the one hand, the transformation of these configurations is important in biological [56, 57] and chemical processes [58, 59], and on the other hand, it is very important for engineering materials. Because this conversion can help to control the electromagnetic polarization of the light and also improve the mechanical properties. For example, materials with different handedness have different transmission effects on circularly polarized electromagnetic waves. It is extremely valuable to pay attention to this feature in the construction of light dividing devices and detectors. Various external stimuli can induce this transformation in natural or synthetic chiral materials. However, the chirality transfer coupled with the control of structural reconfiguration is a very difficult and complex task. For example, with a photoactive medium, chirality can be changed without any structural change in a chiral artificial metamolecule with only light [60]. Therefore, the chirality in multiple scales in a structure is called hierarchical chirality. This type of chirality is very useful for the synthesis and properties of materials, for example, to control symmetry in morphogenesis and increase mechanical properties [61, 62]. Eventually, both chirality transfer and hierarchical chirality provide effective ways to tune the mechanical, optical, or electromagnetic properties of a material.

7. Conclusion and future perspective

Amphiphiles, especially surfactants, due to their unique structure, have been able to make significant progress in industry, various fields of medicine and nanoscience. Since these materials can create aggregated structures in addition to reducing surface/interfacial tension by controlling concentration, structural design and controlling environmental conditions, they can be used as templates or drug carriers. Also, these compounds can be used for household cleaning and other industrial applications, and it is predicted that with the increase in the world population along with industrialization and the increase in living standards in developing countries, the demand for the production of these compounds will increase sharply. On the one hand, the widespread use of these compounds and on the other hand their chemical structure creates environmental concerns regarding their toxicity to mammals, aquatic animals and their accumulation in the soil. Therefore, considering that these materials should be used in some industries because of their unique structure, the design of functional surfactants with less risk for the environment is not far from expected.

The purpose of this chapter is to introduce the structure and assembly behavior of surfactants, trying to find a way so that if we cannot eliminate the use of surfactants in a particular industry due to more disadvantages than its benefits, at least we can reduce the amount of disadvantages by changing its structure. On the other hand, with detailed information about their structure and its effect on their properties, we can suggest the best surfactant with the highest efficiency and minimum environmental and economic harm for a specific application. For example, in 1970, the use of phosphate detergents was banned because phosphate is a nutrient for algae. The entry of the waste of these detergents into the water led to the excessive growth of algae and as a result, the reduction of oxygen in the water, followed by the death of aquatic animals and fish [63]. Since substitutes for these surfactants were found, especially in the current situation, using natural surfactants, so removing or limiting phosphate surfactants will not create a problem in the detergent industry or other industries. But another group of surfactants with chains that have fluorine atoms in their structure, such as perfluorooctane sulfonates (PFOS) and perfluorooctanoic acid (PFOA), are toxic to mammals and are suspected of causing cancer in these organisms [64]. The prohibition on the use of these types of surfactants also forced manufacturers to produce fluorinated surfactants with shorter linear chains. Later studies showed that these types of surfactants have the same performance or in some cases better than the previous surfactants, but their toxicity is less. In some industries, these fluorinated surfactants are not yet easily replaceable because there are no high-performance alternatives. Fluorinated surfactants usually have lower surface tension with excellent wetting ability [65]. Therefore, in order to reduce their toxicity, we should seek to change their structure.

In general, it can be said that with the advancement of computational methods along with complex experimental methods such as X-ray or neutron scattering, nuclear magnetic resonance and electron transfer scanning, the surfactant structure can be designed according to the type of application. In other words, considering that the chemical and thermodynamic laws governing the structures of aggregates formed by surfactants are well understood and documented and the related theories are well advanced, it is possible to use the results of experimental and computational studies and their simultaneous application to design different structures. Surfactants are used for their application in various fields such as nanomaterials, drug delivery, energy efficiency and environmental cleaning.

Author details

Beheshteh Sohrabi
Department of Chemistry, Surface Chemistry Research Laboratory, Iran University of Science and Technology, Tehran, Iran

*Address all correspondence to: sohrabi_b@yahoo.com

References

[1] Israelachvili JN, Mitchell DJ, Ninham BW. Theory of self-assembly of hydrocarbon amphiphiles into micelles and bilayers. Journal of the Chemical Society, Faraday Transactions 2: Molecular and Chemical Physics. 1976; **72**:1525-1568

[2] Kang C, Honciuc A. Self-assembly of Janus nanoparticles into transformable suprastructures. The Journal of Physical Chemistry Letters. 2018;**9**(6):1415-1421

[3] Kang C, Honciuc A. Influence of geometries on the assembly of snowman-shaped Janus nanoparticles. ACS Nano. 2018;**12**(4):3741-3750

[4] Wu D, Chew JW, Honciuc A. Polarity reversal in homologous series of surfactant-free Janus nanoparticles: Toward the next generation of amphiphiles. Langmuir. 2016;**32**(25): 6376-6386

[5] Spitz L. Sodeopec: Soaps, Detergents, Oleochemicals and Personal Care Products. AOCS Publishing; 2004

[6] Sakellariou E. Southern Italy in the late middle ages: Demographic, institutional and economic change in the Kingdom of Naples, c. 1440-c. 1530, Brill. 2011

[7] Behler A, Biermann M, Hill K, Raths H-C, Victor M, Uphues G, et al. Industrial surfactant syntheses. Surfactant Science Series. 2001. pp. 1-44

[8] Palmer LC, Stupp SI. Molecular self-assembly into one-dimensional nanostructures. Accounts of Chemical Research. 2008;**41**(12):1674-1684

[9] Reynhout IC, Cornelissen JJ, Nolte RJ. Synthesis of Polymer– Biohybrids: From small to giant surfactants. Accounts of Chemical Research. 2009;**42**(6):681-692

[10] Yu X, Zhong S, Li X, Tu Y, Yang S, Van Horn RM, et al. A giant surfactant of Polystyrene–(Carboxylic acid-functionalized polyhedral oligomeric silsesquioxane) amphiphile with highly stretched polystyrene tails in micellar assemblies. Journal of the American Chemical Society. 2010;**132**(47): 16741-16744

[11] Zhang J, Song Y-F, Cronin L, Liu T. Self-assembly of organic– Inorganic hybrid amphiphilic surfactants with large polyoxometalates as polar head groups. Journal of the American Chemical Society. 2008;**130**(44):14408-14409

[12] Jeon YJ, Bharadwaj PK, Choi S, Lee JW, Kim K. Supramolecular amphiphiles: Spontaneous formation of vesicles triggered by formation of a charge-transfer complex in a host. Angewandte Chemie. 2002;**114**(23):4654-4656

[13] Versluis F, Tomatsu I, Kehr S, Fregonese C, Tepper AW, Stuart MC, et al. Shape and release control of a peptide decorated vesicle through pH sensitive orthogonal supramolecular interactions. Journal of the American Chemical Society. 2009;**131**(37):13186-13187

[14] Wang C, Wang Z, Zhang X. Superamphiphiles as building blocks for supramolecular engineering: Towards functional materials and surfaces. Small. 2011;7(10):1379-1383

[15] Zhang X, Wang C. Supramolecular amphiphiles. Chemical Society Reviews. 2011;**40**(1):94-101

[16] Epand RM. The Amphipathic Helix. CRC Press; 1993

[17] Li J, Nowak P, Otto S. Dynamic combinatorial libraries: From exploring

molecular recognition to systems chemistry. Journal of the American Chemical Society. 2013;**135**(25): 9222-9239

[18] Minkenberg CB, Florusse L, Eelkema R, Koper GJ, van Esch JH. Triggered self-assembly of simple dynamic covalent surfactants. Journal of the American Chemical Society. 2009;**131**(32): 11274-11275

[19] Li D, Yin P, Liu T. Supramolecular architectures assembled from amphiphilic hybrid polyoxometalates. Dalton Transactions. 2012;**41**(10):2853-2861

[20] Chan GG, Koch CM, Connors LH. Blood proteomic profiling in inherited (ATTRm) and acquired (ATTRwt) forms of transthyretin-associated cardiac amyloidosis. Journal of Proteome Research. 2017;**16**(4):1659-1668

[21] Poggi E, Gohy J-F. Janus particles: From synthesis to application. Colloid and Polymer Science. 2017;**295**(11): 2083-2108

[22] Yang Z, Aarts DA, Chen Y, Jiang S, Hammond P, Kretzschmar I, et al. Janus Particle Synthesis, Self-assembly and Applications. Royal Society of Chemistry; 2012

[23] Yi Y, Sanchez L, Gao Y, Lee K, Yu Y. Interrogating cellular functions with designer Janus particles. Chemistry of Materials. 2017;**29**(4):1448-1460

[24] Zhang Q, Dong R, Chang X, Ren B, Tong Z. Spiropyran-decorated SiO2–Pt janus micromotor: Preparation and light-induced dynamic self-assembly and disassembly. ACS Applied Materials & Interfaces. 2015;**7**(44):24585-24591

[25] Sohrabi B, Gharibi H, Javadian S, Hashemianzadeh M. A new model to study the phase transition from microstructures to nanostructures in ionic/ionic surfactants mixture. The Journal of Physical Chemistry B. 2007; **111**(34):10069-10078

[26] Tanford C. Micelle shape and size. The Journal of Physical Chemistry. 1972; **76**(21):3020-3024

[27] Degiorgio V, Corti M. Physics of Amphiphiles–Micelles, Vesicles, and Microemulsions. Varenna on Lake Como, Villa Monastero: Elsevier Science Limited; 19-29 July 1983, 1985

[28] Israelachvili JN. Intermolecular and Surface Forces. Academic Press; 2011

[29] Zhang J, Li X, Li X. Stimuli-triggered structural engineering of synthetic and biological polymeric assemblies. Progress in Polymer Science. 2012;**37**(8): 1130-1176

[30] Vinson P, Bellare J, Davis H, Miller W, Scriven L. Direct imaging of surfactant micelles, vesicles, discs, and ripple phase structures by cryo-transmission electron microscopy. Journal of Colloid and Interface Science. 1991;**142**(1):74-91

[31] Ziserman L, Abezgauz L, Ramon O, Raghavan SR, Danino D. Origins of the viscosity peak in wormlike micellar solutions. 1. Mixed catanionic surfactants. A cryo-transmission electron microscopy study. Langmuir. 2009;**25**(18):10483-10489

[32] Bockris JOM, Reddy AK, Gamboa-Aldeco ME. Modern Electrochemistry 2B: Electrodics in Chemistry, Engineering, Biology and Environmental Science. Springer Science & Business Media; 1998

[33] Moallemi M, Sohrabi B, Fazeli S. Electrolyte effect on adsorption and the

phase transition from microstructures to nanostructures in ionic/ionic surfactants mixture. Journal of Colloid and Interface Science. 2011;**361**(1):159-169

[34] Nakama Y. Surfactants. Cosmetic science and technology, theoretical principles and applications. 2017. pp. 231-244

[35] Tadros TF. Control of the properties of suspensions. Colloids and Surfaces. 1986;**18**(2–4):137-173

[36] Derjaguin BV. Acta Physicochimica U.R.S.S. Vol. 14. 1941

[37] Verwey EJW, Overbeek JTG. Theory of the stability of lyophobic colloids. Journal of Colloid Science. 1955;**10**(2): 224-225

[38] Tanford C. The Hydrophobic Effect. New York: Wiley, Interscience;

[39] Edsall JT, McKenzie HA. Water and proteins. II. The location and dynamics of water in protein systems and its relation to their stability and properties. Advances in Biophysics. 1983;**16**:53-183

[40] Kawasaki H, Maeda H. FT-IR study on hydrogen bonds between the headgroups of dodecyldimethylamine oxide hemihydrochloride. Langmuir. 2001;**17**(7):2278-2281

[41] Wu LQ, Xu Y, Hou T, Jia JG, Huang XD, Weng GG, et al. Controllable macroscopic chirality of coordination polymers through pH and anion-mediated weak interactions. Chemistry–A European Journal. 2021;**27**(67): 16722-16734

[42] Park KS, Xue Z, Patel BB, An H, Kwok JJ, Kafle P, et al. Chiral emergence in multistep hierarchical assembly of achiral conjugated polymers. Nature Communications. 2022;**13**(1):1-14

[43] Nagarajan R. Molecular packing parameter and surfactant self-assembly: The neglected role of the surfactant tail. Langmuir. 2002;**18**(1):31-38

[44] Sang Y, Liu M. Hierarchical self-assembly into chiral nanostructures. Chemical Science. 2022;**13**(3):633-656

[45] Keene FR. Chirality in Supramolecular Assemblies: Causes and consequences. John Wiley & Sons; 2016

[46] Selinger JV, Spector MS, Schnur JM. Theory of Self-assembled Tubules and Helical Ribbons. ACS Publications; 2001. pp. 7157-7169

[47] Nyrkova I, Semenov A. Twisted surfactant structures: An advanced theoretical model. Soft Matter. 2010;**6** (3):501-516

[48] Selinger RL, Selinger JV, Malanoski AP, Schnur JM. Shape selection in chiral self-assembly. Physical Review Letters. 2004;**93**(15):158103

[49] Wu Z, Yan Y, Huang J. Advanced molecular self-assemblies facilitated by simple molecules. Langmuir. 2014;**30** (48):14375-14384

[50] Batys P, Morga M, Bonarek P, Sammalkorpi M. pH-induced changes in Polypeptide conformation: Force-field comparison with experimental validation. The Journal of Physical Chemistry B. 2020;**124**(14):2961-2972

[51] Feng H, Lv W, Ma J, Chang W, Chen Y, Wang J. Helical structures with switchable and hierarchical chirality. Applied Physics Letters. 2020;**116**(19): 194102

[52] Gao PX, Ding Y, Mai W, Hughes WL, Lao C, Wang ZL. Conversion of zinc oxide nanobelts into

superlattice-structured nanohelices. Science. 2005;**309**(5741):1700-1704

[53] Huang C, Feng Y, Zhao J, Wang Z, Jiang T. Asymmetric electromagnetic wave transmission of linear polarization via polarization conversion through chiral metamaterial structures. Physical Review B. 2012;**85**(19):195131

[54] Pendry J. A chiral route to negative refraction. Science. 2004;**306**(5700): 1353-1355

[55] Wang J-S, Feng X-Q, Xu J, Qin Q-H, Yu S-W. Chirality transfer from molecular to morphological scales in quasi-one-dimensional nanomaterials: A continuum model. Journal of Computational and Theoretical Nanoscience. 2011;**8**(7):1278-1287

[56] Lan X, Liu T, Wang Z, Govorov AO, Yan H, Liu Y. DNA-guided plasmonic helix with switchable chirality. Journal of the American Chemical Society. 2018; **140**(37):11763-11770

[57] Murata K, Aoki M, Suzuki T, Harada T, Kawabata H, Komori T, et al. Thermal and light control of the sol-gel phase transition in cholesterol-based organic gels. Novel helical aggregation modes as detected by circular dichroism and electron microscopic observation. Journal of the American Chemical Society. 1994; **116**(15):6664-6676

[58] de Jong JJ, Lucas LN, Kellogg RM, van Esch JH, Feringa BL. Reversible optical transcription of supramolecular chirality into molecular chirality. Science. 2004;**304**(5668):278-281

[59] Suda M, Thathong Y, Promarak V, Kojima H, Nakamura M, Shiraogawa T, et al. Light-driven molecular switch for reconfigurable spin filters. Nature Communications. 2019;**10**(1):1-7

[60] Zhang S, Zhou J, Park Y-S, Rho J, Singh R, Nam S, et al. Photoinduced handedness switching in terahertz chiral metamolecules. Nature Communications. 2012;**3**(1):1-7

[61] Pokroy B, Kang SH, Mahadevan L, Aizenberg J. Self-organization of a mesoscale bristle into ordered, hierarchical helical assemblies. Science. 2009;**323**(5911):237-240

[62] Zhao Z-L, Zhao H-P, Wang J-S, Zhang Z, Feng X-Q. Mechanical properties of carbon nanotube ropes with hierarchical helical structures. Journal of the Mechanics and Physics of Solids. 2014;**71**:64-83

[63] Chaturvedi V, Kumar A. Toxicity of sodium dodecyl sulfate in fishes and animals. A review. International Journal of Applied Biology and Pharmaceutical Technology. 2010;**1**:630-633

[64] Lau C, Anitole K, Hodes C, Lai D, Pfahles-Hutchens A, Seed J. Perfluoroalkyl acids: A review of monitoring and toxicological findings. Toxicological Sciences. 2007;**99**(2): 366-394

[65] Kissa E. Fluorinated Surfactants and Repellents. CRC Press; 2001

Chapter 4

Electrocatalytic Self-Assembled Nanoarchitectonics for Clean Energy Conversion Applications

Ingrid Ponce, José H. Zagal and Ana María Méndez-Torres

Abstract

The general trends in the construction of highly active electrode devices are focused on the science of materials. These are useful for developing 2D nanostructured electrodes, with well-defined active sites, which are excellent approaches for understanding the fundamentals of electrocatalytic reactions. Here we present an overview of the experimental self-assembled molecular catalyst configurations to develop excellent electrode materials containing molecular catalysts for energy conversion device applications. First, by applying well-known reactivity descriptors for electrocatalysis, nanoarchitectonics, and the self-assembled concept, we summarize the main molecular building blocks to achieve a technology system for arranging by a rational design, nanoscale structural units configuration that promotes electrocatalytic reactions such as oxygen reaction reduction (ORR) and water-splitting reactions. We focus the discussion on the MN4 molecular catalyst linked to electrode surfaces with the help of the axial blocks, bio-inspired self-assembled approaches such as biomimetic models of metalloenzymes active sites, and molybdenum sulfide clusters for hydrogen evolution reaction (HER). We briefly discuss the advantages of developing host-guest self-assembled molecular catalyst systems based on cyclodextrins anchored to electrodes to get well-defined active sites with local environment control.

Keywords: nanostructured electrode, molecular catalyst, oxygen reduction reaction, water splitting, hydrogen evolution reaction

1. Introduction

Climate change is one of the worse problems humanity is facing and despite different measures that different countries are taken to counteract this problem but the situation is not improving. One of the main causes of global warming is the widespread use of fossil fuels, among other factors like cattle farming. Thus, there is an urgent need to find alternative non-polluting energy sources i.e., renewable sources. Among these new energy sources, electrochemical systems appear to be very interesting. For example, a water electrolyzer uses clean electricity to produce high uruty electrolytic hydrogen which can be used as a fuel that upon combustion is pollution free as it forms water. Hydrogen can also be used in a fuel cell producing electricity with high efficiency [1, 2]. For optimum operation of electrolyzers and H_2/O_2 fuel cells, it is necessary to develop highly active catalytic materials to be used in all different electrodes [3, 4]. The development and optimization of these systems are part of the energy revolution that is taking place in the world (UN 2030 agenda), with particular emphasis on the development of new technologies that work more efficiently to promote the decarbonization of the global energy matrix.

Among the most promising alternative energy devices are electrolyzers, fuel cells, as mentioned above, and rechargeable metal-air batteries [5]. These electrochemical systems generally have two electrodes separated by a polymeric membrane, called anode and cathode (**Figure 1a**). Four fundamental electrochemical reactions occur in them for their operation: hydrogen evolution reaction (HER), oxygen evolution reaction (OER), hydrogen oxidation reaction (HOR), and oxygen reduction reaction (ORR). In the water electrolyzers, the H_2O molecule is splitter to give H_2 and O_2 fuels by applying a potential difference between the cathode (HER) and the anode (OER). On the other hand, fuel cells, use H_2 and O_2 to generate electrical energy through the HOR and ORR reactions that occur at the anode and cathode, respectively **Figure 1a**.

HER

$$2H^+ + 2e^- \rightleftharpoons H_2 \text{ acidic electrolytes} \tag{1}$$

OER

$$2H_2O \rightleftharpoons O_2 + 4H^+ + 4e^- \text{ acidic electrolytes} \tag{2}$$

$$4OH^- \rightleftharpoons 2H_2O + O_2 + 4e^- \text{ neutral or alkaline electrolytes} \tag{3}$$

HOR

$$H_2 \rightleftharpoons 2H^+ + 2e^- \text{ acidic electrolytes} \tag{4}$$

ORR

$$O_2 + 4H^+ + 4e^- \rightleftharpoons 2H_2O \text{ acidic electrolytes} \tag{5}$$

$$O_2 + 2H^+ + 2e^- \rightleftharpoons H_2O_2 \tag{6}$$

$$O_2 + 2H_2O + 4e^- \rightleftharpoons 4OH^- \text{ alkaline electrolytes} \tag{7}$$

$$O_2 + H_2O + 2e^- \rightleftharpoons HO_2^- + OH^- \tag{8}$$

The above reactions occur by multi-electron pathways and, therefore, can proceed via several mechanisms. For example, in acid electrolyte environments, HER follows two elemental reactions, proton adsorption on the active site of electrode surface (M) by the Volmer mechanism, and desorption of the H_2 molecule step by Heyrovsky or Tafel mechanisms. Thus, if the *M-H* interaction is too weak, the *adsorption* step will limit the overall reaction rate, and if the *M-H* interaction is too strong, the *reaction – desorption* step will limit the overall reaction rate [7]. As a result, optimal catalytic surfaces exhibit a Gibbs free energy of electrochemical H-adsorption (ΔGH) close to zero (**Figure 1b**) as Parson proposed in 1958 [8]. On the other hand, HOR proceeds by the same steps but reversely for HER [2, 5].

$$H^{+} + e^{-} + M \rightleftharpoons M - H \text{ Volmer step} \tag{9}$$

$$M - H + H^{+} + e^{-} \rightleftharpoons H_2 + M \text{ Heyrovsky step} \tag{10}$$

$$2M - H \rightleftharpoons H_2 + 2M \text{ Tafel step} \tag{11}$$

In this direction, the ORR can proceed by the four-electron pathway ($4e-/4H+$), which is desirable for fuel cells and air batteries; the two-electron pathway ($2e-/2H+$) promotes H_2O_2 as a product. Therefore, from the energy conversion challenges, an efficient catalyst for ORR is the one that achieves to promotes a selectivity for $4e-/4H+$ *on* $2e-/2H+$.

The HER, HOR, OER, and ORR reactions occur through an internal sphere electron transfer process and involve electrocatalytic phenomena. Thus, their efficiency is directly related to the electrodic materials used as anode and cathode [3, 9, 10]. Therefore, it requires electrodes with specific activity for the reaction involved (**Figure 1b–d**). However, in most known electrode materials, these reactions are slow and require specific catalysts to reach rates compatible with their applicability at the industrial level, which are based on expensive platinum-based materials or precious metal oxides such as IrO_2 or RuO_2 [2, 5, 11]. Noble metals are scarce and expensive, preventing the mass use of these devices. In this context, new alternatives to using inexpensive materials have been proposed. Among them, the graphite materials modified with the MN4 molecular catalysts, such as metallophthalocyanines, metalloporphyrins, and their derivatives complexes of the metal transition (**Figure 1e**).

Graphite electrodes modified by a simple process involving the physical adsorption process of intact MN4 complexes on their surface have made it possible to perform fundamental studies to establish reactivity descriptors for specific reactions such as ORR and, with this, advance in the search for the best catalyst for this reaction through a rational design [9, 10] of the MN4 molecules. However, new experimental strategies and surface characterization techniques have allowed the development of more advanced electrode systems based on nanostructured systems of molecular catalysts as the main active component [6, 12–14]. Thus, one of the fundamental strategies to incorporate a molecular catalyst such, as MN4 complexes, on electrode surfaces is using the self-assembled process, where the molecular catalyst is anchored to the electrode surface using a molecular building block that can establish a link between surface-molecular catalysts. [15] Among these systems, the most prominent are the self-assembled organic monolayers (SAMs) on gold electrode surfaces and their subsequent functionalization with MN4 complexes that can be converted into regular molecular arrays by molecular-level nanoarchitectonics (**Figure 1f**) [16].

The concept of nanoarchitectonics has been proposed as a technology system for arranging nanoscale structural units in a rationally pre-designed configuration for a given function [17]. Thus, in self-assembled molecular catalyst systems on an electrode surface, each functionalized SAM molecule fulfills the function of molecular wire and, at the same time, as an axial ligand of the MN4 complex. This configuration allows the catalyst to be anchored on the surface, permitting the electronic communication between the electrode and molecular catalyst (**Figure 1f**). Therefore, through the self-assembly process, it is possible to obtain nanostructures with a specific functionality on the electrode that has allowed the generation of efficient electrode systems with reproducible activity for specific reactions to advance the understanding of reaction mechanisms achieving to obtain the control of one pathway reaction over other [12, 14]. Moreover, self-assembled nanostructures provide the fundamental

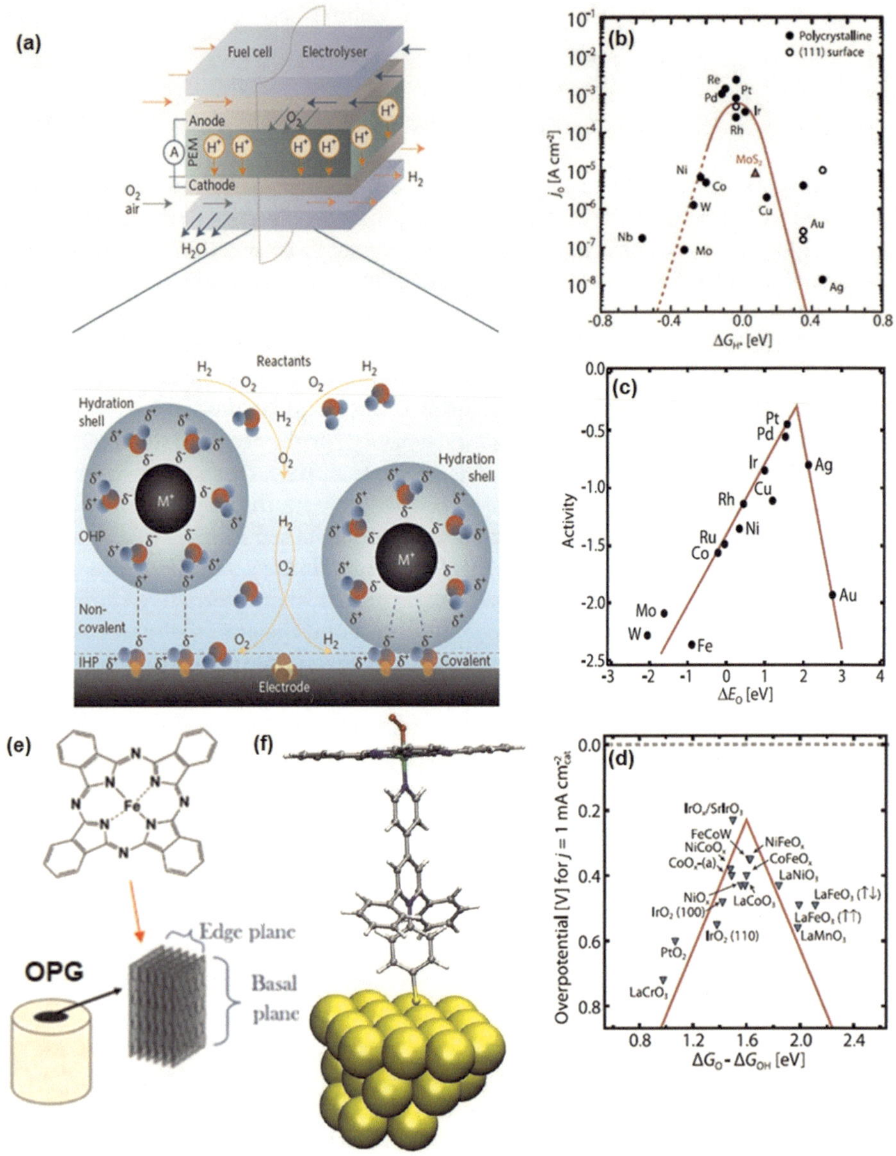

Figure 1.
(a) Scheme of energy conversion devices: Fuel cells and electrolyzer. In the polymer electrolytic membrane (PEM) fuel cell and electrolyzer, the cathode and anode systems are separated by a proton-conducting membrane. Shown on the left is the fuel cell, which is supplied with hydrogen fuel by oxidizing at the anode, producing H^+ and e^-. while, at the cathode, the oxygen in the air reacts with H^+ and e^-. on the right side, an electrolyzer is shown in which the operation is reversed, i.e., at the anode, the water splitting reaction leads to the production of O_2, while the H_2 and electrons at the cathode side are used. To convert H^+ to H_2 reprinted by permission from [springer nature customer service Centre GmbH]: [springer nature] [nature materials] [1] (energy and fuels from electrochemical interfaces, V.R. Stamenkovic, D. Strmcnik, P.P. Lopes, N.M. Markovic), [COPYRIGHT] (2017). (b) HER volcano plot for metals and MoS_2. (c) ORR volcano plot for metals. (d) OER volcano plot for metal oxides. Reprinted by permission from [the American Association for the Advancement of Science AAAS]: [science] [2] (combining theory and experiment in electrocatalysis: Insights into materials design, Z.W. Seh, J. Kibsgaard, C. F. Dickens, I. Chorkendorff, J.K. Nørskov, T.F. Jaramillo), [COPYRIGHT] (2017). (e) Schematic representation of MN4-modified ordinary pyrolytic graphite (OPG) electrode. (f) Self-assembled system made up of electrocatalyst attached by a wire attached to the gold substrate. Reprinted (adapted) with permission from the journal of physical chemistry C [6]. Copyright (2012) American Chemical Society.

basis for constructing supramolecular nano-devices that can be used as active components to build compact and efficient electrodic devices for energy conversion systems.

According to above, the following sections present the fundamental aspects of the construction of molecular catalysts self-assembled to electrode surfaces, which serve as self-assembled supramolecular electrocatalytic devices. For this, it is necessary to consider each molecular building block that, through a rational design, allows the formation of nanostructures guided by molecular self-organization processes on electrode surfaces; in addition, the optimization of electrocatalytic properties of the molecular catalyst for a specific reaction of energetic interest. We will focus mainly on ORR and HER processes, ending with an overview of the new perspectives to consider.

2. Modified electrodes with MN4 complexes as electrocatalytic building molecular blocks

The concept of modified electrodes refers to those electrodes whose surface has been altered by incorporating compounds that can act as mediators in a charge transfer reaction (**Figure 2a-c**) [18]. These compounds can be metal complexes, organometallic complexes, conductive polymeric films, metallic atoms, and self-assembled molecules confined, deposited, or adsorbed on the electrode surface. The last strategy is the first approach to incorporate, by a simple method, new active sites on the electrode surface by the physical adsorption process of molecular catalyst; in this process, the molecules are attached to the surface through weak bonds such as van der Waals forces and π-staking interactions, among others. Thus, the modified electrode can present electrocatalytic activity for a given reaction. In this case, the electrocatalytic activity is greater than that of the unmodified electrode (**Figure 2f-g**) [18].

MN4-type macrocyclic complexes are well-known to exhibit electrocatalytic activity for different reactions of interest, including ORR. When incorporated into the electrode surfaces, these act as charge mediators, creating new active sites on the surface and reducing the overpotential needed to initiate the electrochemical reaction (**Figure 2f-g**). In this direction, the studies have focused on exploring the reactivity descriptors for a specific reaction such as ORR in ordinary pyrolytic graphite (OPG) and glassy carbon electrodes modified directly with MN4. The importance of these electrode surfaces modified with MN4 lies in the fact that, through certain experimental conditions, it is possible to achieve a high degree of control of the surface in terms of its structure and reactivity, being able to establish and control its reactivity and specificity. Most MN4 complexes studied have involved transition metal phthalocyanines (MPcs), metal porphyrins (MP), and similar complexes based on biomolecules (e.g., Vit B12, Cytochrome c, and multicopper oxidase) adsorbed directly on carbon, and pyrolytic graphite materials, among others [11]. This is due to the variety of electrochemical reactions catalyzed by these systems.

In MN4 complexes, the metal center acts as an active site; in addition, they have a flat and rigid structure with a highly delocalized electronic system (**Figure 2a-c**). These can undergo very fast reversible redox processes with a minimum of molecular reorganization. In addition, its redox properties can be tunned using electron donor or acceptor substituents in the cyclic ligand, thus, regulating the electron density on the metal center. The redox processes of MN4 are closely related to the energies of the frontier orbitals, which are precisely those involved in the interaction with the reacting molecule. [6, 12–14]. Among the most studied MN4 complexes are phthalocyanines and porphyrins. However, new highly active complexes for ORR have

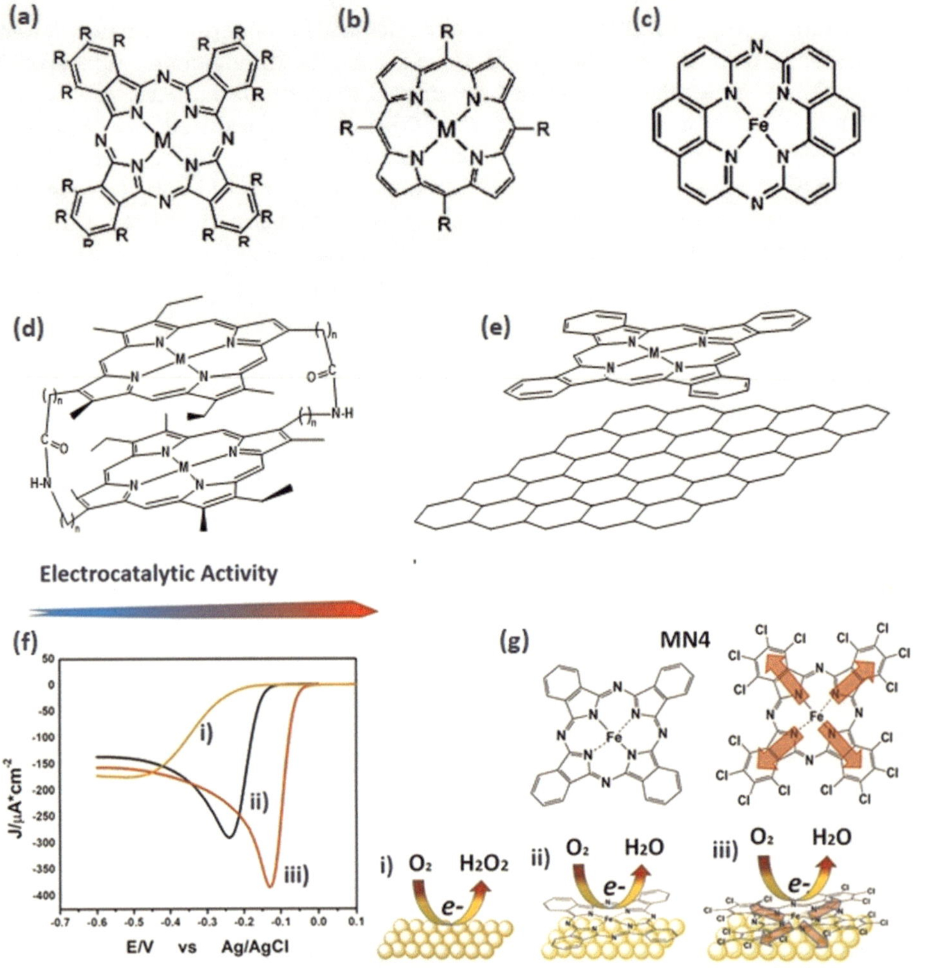

Figure 2.
Examples of commonly used macrocyclic compounds: (a) Metalphthalocyanine. (b) Metalloporphyrin, (c) phenanthroline derivative, (d) dicobalt cofacial porphyrin M = Co, and (e) cobalt phthalocyanine on graphite M = Co. f) Voltammetric response for O_2 reduction for a clean and/or modified (111) gold electrode with different configurations containing MN4 complexes. (g) MN4 complexes: Iron phthalocyanine (FePc) and hexadecachlorinated phthalocyanine (16(Cl)FePc). ORR schemes in the unmodified gold electrode (i) and modified with FePc (ii) and 16(Cl)FePc (iii) measurements made in a 0.1 M NaOH solution saturated with O_2, dE/dt = 50 mV/s. reprinted (adapted) with permission from ACS catalysis [12]. Copyright (2018) American Chemical Society.

recently emerged, such as Fe phenanthroline derivatives, which mimic the M-N4 site **(Figure 2c)** [17].

In these systems, reactivity descriptors have been found for ORR, such as the number of "d" electrons in the metal center, the donor-acceptor intermolecular hardness, the binding energy between the metal and oxygen, and the redox potential $M^{n+}/M^{(n-1)+}$; the latter two being closely related [9]. These reactivity descriptors allow the synthesis of new and optimized molecular catalysts based on these complexes. In iron phthalocyanine, it has been shown that by incorporating electron-withdrawing substituents at the periphery of the ligand macrocycle, the electrocatalytic activity for ORR increases. Electron withdrawing ligands would promote a harder active site, tuning a better *Fe-*O_2 interaction since the O_2 molecule is a hard molecule; the Sabatier principle supports this effect [12]. The control binding energies of the reacting molecule on an active site of the electrode surface are the key to designing materials with improved performance [2]. In this direction, multinuclear complexes onto graphite electrodes are selective for the $4e^-/4H^+$ reduction for ORR over the $2e^-/2H^+$ pathway; thus, a high selectivity has been obtained for dicobalt cofacial porphyrin and related complexes (**Figure** 2d-e) [19]. In those models, the two metal centers help to improve the electrocatalytic performance due to the interaction of the O_2 molecule linked to the multinuclear active site [19].

2.1 Molecular catalyst self-assembled to electrode surface: a experimental platform to obtain 2D nanomaterials for clean energy conversion applications

MN4 molecular catalysts and derivatives have functional electrocatalytic properties to be used as active components in constructing nanostructured electrode surfaces for energy conversion systems. As mentioned in the previous section, their electrocatalytic properties can be tuned by the action of the substituents on the cyclic ligand (**Figure 2g**). Moreover, molecular catalysts such as metal phthalocyanines of Fe, Co, Mn (FePc, CoPc, MnPc) can accept axial extraplanar ligands directly coordinate with the metal center. Exploiting this ability, the complexes can be anchored to electrode surfaces using an axial ligand by a bottom-up construction procedure to obtain well-defined nanostructures [12, 14].

One strategy to immobilize MN4 complexes on electrode surfaces is functionalizing self-assembled organic monolayers, SAMs; among the most used surfaces to carry out this process are gold surfaces.

SAMs are a set of molecular building blocks spontaneously formed in a 2D arrangement (D = dimension) on an electrode surface, forming films of nanometric height [15, 20]. The building blocks that form the SAMs have a well-defined structure to give them a certain chemical functionality, which consists of a "head" and a "terminal" group, which binds strongly to the substrate at one extreme and, another extreme, limit toward the outer surface in contact with the electrolytic solution, respectively. Connecting the head and terminal groups is carried out by a spacer or backbone of variable structure and length [15]. Thus, if SAMs on an electrode surface are considered 2D nanomaterials according to their dimensions, the functionalization of the SAMs with molecular catalyst represents an excellent platform for the construction of electrocatalytic patterned surfaces to apply to the fabrication of compact and efficient electrodic devices (**Figure 3a**).

In the self-assembled arrays, there is a synergic contribution at the molecular level by a self-controlled organization. Therefore, by a predesignated configuration, self-assembled FePc among others, and related metal complexes systems have been developed on gold surfaces for ORR. In this surface configuration, at the nanoscale, each electrocatalytic nanostructure has a molecular catalyst with an active site conformed by a five-coordinated metal center forming an umbrella-*like* configuration (**Figure 3b**). As mentioned above, axially coordinated molecular catalysts present higher performance activity in comparison with the adsorbed systems. So, the axial ligand apart from serving as a molecular anchor is not innocent and affects the electron density on the metal center. Thus, in well-established electrocatalytic molecular platforms involving this type of self-assembled nanoarchitectonic for FePc and related complexes, the activity for ORR is enhanced by the axial ligand action compared to the FePc directly adsorbed on the electrode surface. Similar self-assembled

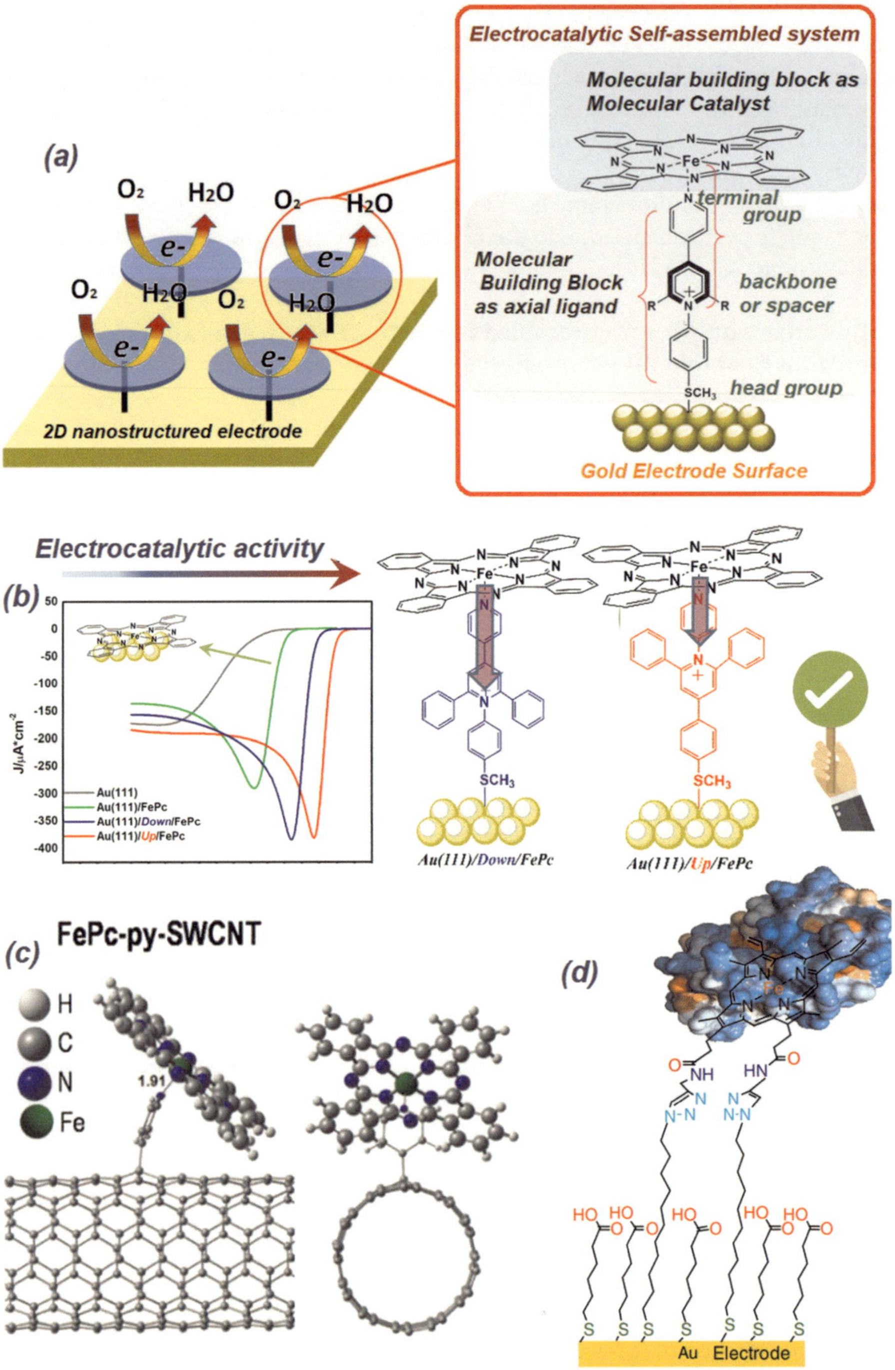

Figure 3.

(a) Schematic representation of 3D nanostructured gold electrode obtained by the bottom-up construction of self-assembled monolayers functionalized with FePc. (b) Lineal sweet voltammetry for ORR on Au(111) modified with different configurations of pyridiniums molecules with a electron-withdrawing groups into molecular backbone. Performed on 0.1 M NaOH O_2 saturated, dE/dt = 0.05 V s^{-1}. Reprinted (adapted) with permission from Electrochimica Acta [14]. Copyright (2018) Elsevier. (c) Optimized molecular structures of FePc-py-SWCNT. Reprinted (adapted) with permission from Electrochimica Acta [21]. Copyright (2021) Elsevier. (d) Construction of the electrode bearing the biosynthetic model. Adapted from ***Figure 2*** *in ref. [22].*

systems have been carried out on single-walled carbon nanotubes, SWCNTs, where the axial ligand corresponds to a building block anchored to the carbon surface by a covalent bond, SWCNTs-Py systems (**Figure 3c**). However, FePc is self-assembled in SWCNTs-Py systems by spontaneous interaction between Fe and the axial ligand available in SWCNTs-Py systems [21].

In a broad scene, three components are fundamental for obtaining self-assembled molecular catalyst arrays:

1. The electrode surface, which is the crucial material for electrochemical systems represents the source and sink of the electrons. The nature of the surface (such as graphite, gold, and Ag) depends on the type of molecular building block and, therefore, the kind of interactions that drive the bottom-up self-assembly process (e.g., van der Waals interaction, π-stacking, among others).

2. A functional molecular building block to act as an axial ligand to the metal center of the molecular catalyst. The general rational design of this building block must consider incorporating an "anchor" functional group at one end, which has a chemical affinity to bind to the electrode surface (e.g., SH and -SCH_3 for gold and pyrene for graphite surfaces). Also, a coordinating functional group binds to the metal center of an MN4 complex (e.g., Pyridine and NH_2, it contains atoms that can donate a pair of electrons to the metal center). Moreover, the rational design must be considered the reactivity descriptors for obtaining a molecular backbone where it aids in exploiting the crucial factors to improve the activity of the self-assembled systems for an interesting reaction. Thus, for ORR carried out on FePc, the activity is increased when the molecular catalyst is anchored to the gold surface by a building block with electron-withdrawing groups into the molecular backbone (**Figure 3b**). This effect can be related to the $donor - acceptor$ Fe – O_2 intermolecular hardness. The electron-withdrawing groups, although they decrease the electron density on the metal center, favor the donor–acceptor electronic coupling, and increase the catalytic activity of FePc for the ORR. In similar highly catalytic systems, *FePc-Py-SWCNTs* the presence of pyridine neutral axial ligand decreases the electron density on the Fe site compared with that of FePc adsorbed on the SWCNTs surface [12, 21].

3. An electrocatalytic molecular building block, such as MN4 complexes. In the case of incorporating the complex to the surface in an umbrella-*like* configuration (**Figure 3b**), it is required that the central metal (Fe, Co, Mn) can accept extraplanar ligands. In addition, as mentioned in the previous point and Section 2, a variety of reactivity descriptors have been proposed for the MN4 complexes for the ORR. In addition, as mentioned in the previous point and Section 2, various reactivity descriptors have been proposed for the MN4 complexes for the ORR. Among them, the M(III)/(II) formal potential of the catalysts is an excellent reactivity predictor that requires an accurate determination of the surface modified with MN4 complexes. Furthermore, these complexes present reversible redox peaks that have been assigned to M(II)/(I) and M(III)/(II) processes supported by spectroscopic evidence [12].

The bioinspired nanostructured electrodes are an excellent experimental platform for mimi kg the efficiency and selectivity of natural metalloenzymes (**Figure 4**a). Thus, the electrode bearing the biosynthetic model has been constructed for electrochemical ORR by "exploiting" the bio-architecture of the active site configuration. As mentioned earlier, multinuclear complexes are selective for the four-electron reduction for ORR over the two-electron pathway. Thus, the biomimetic models of Cytochrome c oxidase (CcO) have been carried out for ORR, where the active site corresponds to a heteronuclear complex (Fe heme, a distal copper, and tyrosine residues) anchored to a gold electrode surface by the action of imidazole ligand linked to alkanethiols SAMs on gold electrodes. In those systems, the synergic contribution of axial ligand and multinuclear active sites helps to improve the electrocatalytic performance for ORR due to the improved interaction of the O_2 molecule linked to the multinuclear active site [11, 23]. The same principle is fulfilled in self-assembled monolayers functionalized by covalently attached mutant myoglobin based on the biosynthetic model of CcO to carry out ORR where both the distal Cu and the redox-active tyrosine residue can help both electron and proton transfer during O_2 reduction [22].

In search of functional molecular building blocks to obtain a low-cost catalyst for the hydrogen evolution reaction (HER), sulfur-rich transition metal clusters have been used to modify graphite and gold electrodic surfaces due to containing groups that resemble the active core groups in redox metalloenzymes, such as iron–sulfur cluster-based enzymes as well as Mo-enzymes; in truth, many processes in nature also rely on metal sulfide clusters [8, 24]. Thus, self-assembled inorganic monolayers of incomplete cubane type $[Mo_3S_4]^{4+}$ discrete clusters have been developed on the gold electrode surface by direct interaction between atom sulfur linked to Au(111) surfaces (**Figure 4**b). The $[Mo_3S_4]^{4+}$ unit comprises three molybdenum atoms in the oxidation state +4 positioned in a triangular plane capped by a μ_3-S^{2-} entity. The four sulfur atoms in the cluster are in the reduced state −2 like –SH "head" functional group in organic thiols [8]. Therefore, in the self-assembled $[Mo_3S_4]^{4+}$/Au(111) systems, the most stable configuration would be three sulfur atoms acting as linkers to gold and one sulfur atom exposed between the solution electrode interface (**Figure 4b**). This

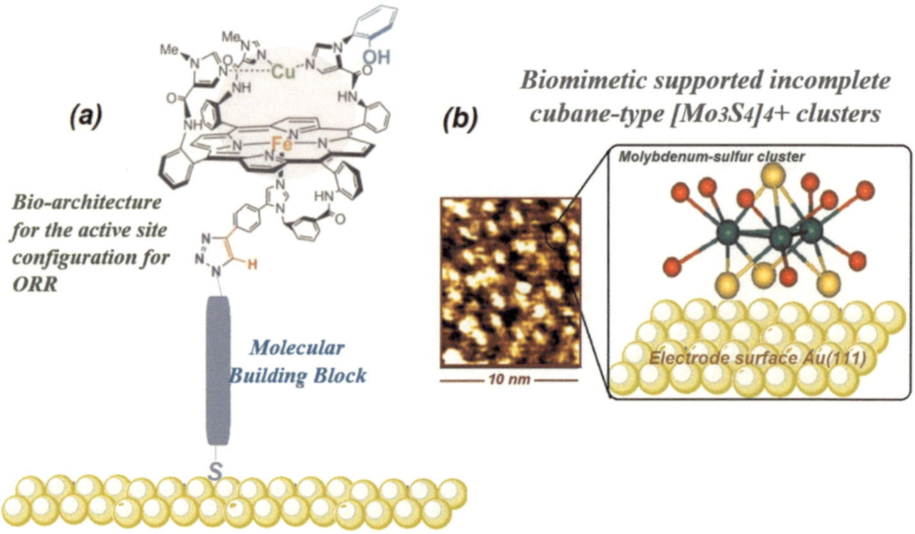

Figure 4.
(a) Schematic representation of SAMs functionalized with synthetic CcO model. Adapted from Figure 3 in ref. [23]. (b) in situ *STM images of $[Mo_3S_4]^{4+}$ monolayer on Au(111) in 0.1 M $HClO_4$ 10x10 nm^2. Adapted from Figure 7 in ref. [24].*

configuration has been characterized by different experimental techniques such as cyclic and lineal voltammetry, XPS, and in-situ scanning tunneling microscopy, STM [7, 24]. This system works efficiently for electrocatalysis for hydrogen evolution reaction, where the onset potential (the potential at which catalytic current is first observed) is shifted 200 mV, in an acidic environment, to favorable potentials concerning the gold surface without inorganic SAMs [24].

Earlier, we mentioned the three possible reaction steps involved in the mechanism for HER (reactions *9*, *10*, and *11*). Whether the reaction proceeds via the *Volmer – Heyrowsky* mechanism or the *Volmer – Tafel* mechanism, the reaction occurs by hydrogen atoms adsorbed on the electrode surface, *M-H* (where M = active site on electrode surface) [7]. On the other hand, these inorganic clusters consist of small molecular units of molybdenum sulfide with undercoordinated sulfur abounding at its surface, and it is generally assumed that hydrogen binds to under coordinated sulfur edge sites. Therefore, this self-assembled discrete unit on electrode surfaces provides an efficient active site for carrying out the HER process, where the overall reaction rate is influenced by the free energy of hydrogen adsorption, ΔGH (**Figure 1b**) [7].

Because molybdenum sulfide clusters have shown great promise for electrocatalysis for HER to generate carbon-free fuel from water, different experimental techniques have been proposed to incorporate them on electrode surfaces. Among these techniques, straightforward methods involving Mo-S clusters are supported on various substrates by simple drop-casting from a specific solution. Thus, the highly oriented pyrolytic graphite (HOPG) electrode surfaces have been modified with $[Mo_3S_4]^{4+}$ molecular clusters by physical adsorption or electrostatic interaction on anodized surfaces. In this configuration, it was possible to obtain a *structure–function* relationship of these electrocatalytic building blocks at the single-molecule level; this latter, by combining electrochemistry and STM experiments. In those systems, the molecule cluster activity was higher than in nonprecious metal HER catalysts, such as Ni, Cu, or W. Similar behavior has shown $[Mo_3S_{13}]^{2-}$ nanoclusters [7]. In this direction, Mo-S represents promising building block entities to obtain electrocatalytic nanostructured electrode surfaces in well-established arrays directed by different self-assembled ways such as quasi-covalent interaction (S-Au), van der Waals interaction, electrostatic interaction (between surface groups and cluster unit), and in-situ ligand substitution to obtain electrocatalytic linked blocks arrays [7].

2.2 Electrocatalysis based on host-guest systems

Recently, a new perspective has emerged in the development of nano-device with potential applications in energy conversion reactions, which is the formation of host-guest systems at the electrode interface [25, 26]. Within the wide range of hosts, molecules are cyclodextrins, calixarenes, and cucurbit[n]urils, among others [27, 28], which are based on the formation of inclusion complexes through non-covalent interactions [29].

Cyclodextrins (CDs) are functional building blocks corresponding to a family of macrocyclic oligosaccharides composed of D-(+)-glucopyranose subunits linked through an α-1,4-glucosidic bond. The most common CDs are composed of 6, 7, or 8 glucopyranose subunits and are called α -CD, β-CD, and γ-CD, respectively [30]. The CDs exhibit two types of –OH groups at three different positions: primary –OH groups at the C_6 position and secondary –OH groups at C_2 and C_3 positions. These –OH groups are located on the edges providing a hydrophilic exterior, whereas C–H units at the inside create a hydrophobic [31, 32]. Among the native CDs, the most widely used due to its greater accessibility and functionalization of its chemical structure is β-CD. This CD can form inclusion complexes with a series of aromatic and heterocyclic compounds [33]. In this direction, as an experimental platform, these entities are suitable building blocks to obtain well-structured arrays in 2D conformation on electrodes surfaces to study, at the single-molecule level, the influence of the local environment hydrophobicity/hydrophilicity on the electron-transfer process, across the self-assembled molecular catalyst for ORR, HER, OER, and HOR [31]. Therefore, the host-guest self-assembled molecular catalyst could provide a directed via to develop a fundamental understanding of the intrinsic activity of catalytic sites in the interface's electrode solutions, which is necessary to design highly active surfaces whit well-defined active sites.

Komatsu. et al. reported the formation of a host-guest system building from α-CDs units and porphyrin-pyridinium complexes with an iron penta-coordinated metal center. In this system, a stable adduct with an oxygen molecule forms through the direct interaction of Fe-O2, **Figure 5a** [34]. Other host-guest systems between thiolated cyclodextrins (β-CDSH) and iron porphyrin (FeTMPyP) have been developed to promote the ORR via the 4-electrons reduction process [36]. In these self-assembled systems, the MN4 complex is linked to the electrode surface modified with CD by the favorable interaction of the axial ligand into de hydrophobic cavity into the CD. On the other hand, cobalt porphyrins with different substituents on the periphery of the macrocycle rings ((TPPS)Co, (TMPyP)Co, (TPP)Co)) were used to form inclusion complexes with β-CD polymers immobilized on gold to evaluate electrocatalysis of ORR (**Figure 5b**) [35, 37]. This reaction at the electrode coated with a polymer film, hosting the cobalt porphyrin catalyst, involves a 2e-/2H+ pathway resulting in the formation of hydrogen peroxide.

Besides, the host-guest inclusion complexes have been carried out with ruthenium electrocatalysts and per-thiolated β-CDs [31] anchored on gold electrode surfaces. This self-assembled configuration favors the ammonia electrooxidation reaction (AOR) since ammonia (NH_3) shows promise as a renewable hydrogen storage medium, i.e., by electrochemical decomposition for on-site hydrogen generation or by direct use of ammonia as fuel in a fuel cell. This shows that the development of AOR electrocatalysts is essential to obtain a high yield in HER and ORR.

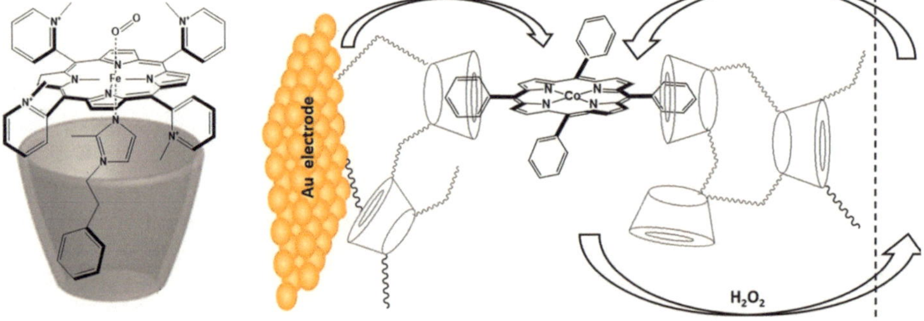

Figure 5.
(a) Schematic representation of possible dominant αααβ structure of the deoxygenated host-guest ensemble. Adapted from reference [34]. (b) Scheme of electrocatalytic dioxygen sensing at the gold electrode coated with the β-CD film containing cobalt porphyrin (Co(TPP)). Adapted from reference [35].

3. Conclusion

We have described the crucial role of rational design in obtaining molecular building blocks as active components for constructing a self-assembled molecular catalyst for the electrocatalysis of the energetic reactions interest, focusing on the ORR and HER process. These reactions occur through an inner-sphere electron transfer process and involve electrocatalytic phenomena; therefore, their efficiency is directly related to the electrodic materials used as cathode. We summarized the reactivity descriptors applied to build self-assembled molecular catalysts to promote electrocatalytic reactions in the central molecular configurations to gain a fundamental understanding of obtaining nanostructured electrode surfaces. We make the description for different experimental approaches to achieve self-assembled arrays with synergic contribution, at the molecular level, by a self-controlled organization. Among this configuration, MN4 molecular catalyst linked to gold electrodes and graphite surfaces by the axial ligands have been developed to improve the activity for ORR compared with catalyst adsorbed directly on electrode surfaces. Furthermore, we described a few bio-inspired self-assembled approaches based on biomimetic architecture to emulate similar metalloenzymes active sites. Finally, we describe the advantages of developing host-guest self-assembled molecular catalyst systems based on cyclodextrins anchored to gold electrodes to obtain highly active surfaces whit well-defined active sites with local environment hydrophobicity/hydrophilicity control on the electron-transfer process.

Acknowledgements

This work has been financed by the Fondecyt Project 1211351. Fondecyt Project and 1221798. Ring Project ACT 192175 and ANID-Fondecyt Postdoctoral 3220728.

Author details

Ingrid Ponce*, José H. Zagal and Ana María Méndez-Torres
Faculty of Chemistry and Biology, Laboratory of Molecular Electronics and Electrocatalysis, Department of Chemistry of the Environment, Department de Chemistry of Materials, University of Santiago de Chile, Santiago, Chile

*Address all correspondence to: ingrid.ponce@usach.cl

References

[1] Stamenkovic VR, Strmcnik D, Lopes PP, Markovic NM. Energy and fuels from electrochemical interfaces. Nature Materials. 2017;**16**(1):57-69

[2] Seh ZW, Kibsgaard J, Dickens CF, Chorkendorff I, Nørskov JK, Jaramillo TF. Combining theory and experiment in electrocatalysis: Insights into materials design. Science. 2017;**355**:6321

[3] Zagal JH, Bedioui F. Electrochemistry of N4 macrocyclic metal complexes. In: Zagal JH, Bedioui F, editors. Electrochemistry of N4 Macrocyclic Metal Complexes: Volume 1: Energy. Second ed. Vol. 1. Cham: Springer International Publishing; 2016. pp. 1-316

[4] Kim C, Dionigi F, Beermann V, Wang X, Möller T, Strasser P. Alloy Nanocatalysts for the electrochemical oxygen reduction (ORR) and the direct electrochemical carbon dioxide reduction reaction (CO 2 RR). Advanced Materials. 2019;**31**(31):1805617

[5] Chia X, Pumera M. Characteristics and performance of two-dimensional materials for electrocatalysis. Nature Catalysis. 2018;**1**(12):909-921

[6] Ponce I, Silva JF, Oñate R, Rezende MC, Paez MA, Zagal JH, et al. Enhancement of the catalytic activity of Fe Phthalocyanine for the reduction of O 2 anchored to Au(111) via conjugated self-assembled monolayers of aromatic thiols As compared to Cu Phthalocyanine. Journal of Physical Chemistry C. 2012;**116**(29):15329-15341

[7] Benck JD, Hellstern TR, Kibsgaard J, Chakthranont P, Jaramillo TF. Catalyzing the hydrogen evolution reaction (HER) with molybdenum sulfide nanomaterials. ACS Catalysis. 2014;**4**(11):3957-3971

[8] Jaramillo TF, Bonde J, Zhang J, Ooi B-L, Andersson K, Ulstrup J, et al. Hydrogen evolution on supported incomplete Cubane-type [Mo 3 S 4] 4+ Electrocatalysts. Journal of Physical Chemistry C. 2008;**112**(45):17492-17498

[9] Zagal JH, Koper MTM. Reactivity descriptors for the activity of molecular MN4 catalysts for the oxygen reduction reaction. Angew Chemie Int Ed. 2016; **55**(47):14510-14521

[10] Zagal HJ, Ponce I, Oñate R. Redox potentials as reactivity descriptors in electrochemistry. In: Redox. IntechOpen; 2020. pp. 1-24

[11] Dey S, Mondal B, Chatterjee S, Rana A, Amanullah S, Dey A. Molecular electrocatalysts for the oxygen reduction reaction. Nature Reviews Chemistry. 2017;**1**(12):0098

[12] Pizarro A, Abarca G, Gutiérrez-Cerón C, Cortés-Arriagada D, Bernardi F, Berrios C, et al. Building Pyridinium molecular wires as axial ligands for tuning the Electrocatalytic activity of iron Phthalocyanines for the oxygen reduction reaction. ACS Catalysis. 2018;**8**(9):8406-8419

[13] Ponce I, Aragonès AC, Darwish N, Pla-Vilanova P, Oñate R, Rezende MC, et al. Building nanoscale molecular wires exploiting Electrocatalytic interactions. Electrochimica Acta. 2015; **179**:611-617

[14] Gutiérrez-Ceron C, Oñate R, Zagal JH, Pizarro A, Silva JF, Castro-Castillo C, et al. Molecular conductance versus inductive effects of axial ligands on the electrocatalytic activity of self-assembled iron phthalocyanines: The oxygen reduction reaction. Electrochimica Acta. 2019;**327**:1-12

[15] Vericat C, Vela ME, Benitez G, Carro P, Salvarezza RC. Self-assembled monolayers of thiols and dithiols on gold: New challenges for a well-known system. Chemical Society Reviews. 2010; **39**(5):1805

[16] Bardosova M, Wagner T. Nanomaterials and Nanoarchitectures. In: Bardosova M, Wagner T, editors. NATO Science for Peace and Security Series C: Environmental Security. Dordrecht: Springer Netherlands; 2015

[17] Ariga K, Li M, Richards G, Hill J. Nanoarchitectonics: A conceptual paradigm for design and synthesis of dimension-controlled functional nanomaterials. Journal of Nanoscience and Nanotechnology. 2011;**11**(1):1-13

[18] Zagal JH, Bedioui F, Gutiérrez S, Alatorre A. Electrocatálisis, aspectos teóricos y experimentales. In: Zagal JH, editor. ELECTROQUÍMICA: VOLTAMETRÍAS SOBRE ELECTRODO SÓLIDO. 1st ed. Santiago: Universidad de Santiago de Chile; 2009. p. 336

[19] Durand RR, Bencosme CS, Collman JP, Anson FC. Mechanistic aspects of the catalytic reduction of dioxygen by cofacial metalloporphyrins. Journal of the American Chemical Society. 1983;**105**(9):2710-2718

[20] Love JC, Estroff LA, Kriebel JK, Nuzzo RG, Whitesides GM. Self-assembled monolayers of thiolates on metals as a form of nanotechnology. Chemical Reviews. 2005;**105**(4): 1103-1170

[21] Oyarzún MP, Silva N, Cortés-Arriagada D, Silva JF, Ponce IO, Flores M, et al. Enhancing the electrocatalytic activity of Fe phthalocyanines for the oxygen reduction reaction by the presence of axial ligands: Pyridine-functionalized single-walled carbon nanotubes. Electrochimica Acta. 2021;**398**:1-13

[22] Mukherjee S, Mukherjee A, Bhagi-Damodaran A, Mukherjee M, Lu Y, Dey A. A biosynthetic model of cytochrome c oxidase as an electrocatalyst for oxygen reduction. Nature Communications. 2015;**6**(1):8467

[23] Collman JP, Devaraj NK, Decréau RA, Yang Y, Yan Y-L, Ebina W, et al. A cytochrome c oxidase model catalyzes oxygen to water reduction under rate-limiting electron flux. Science. 2007;**315**(5818):1565-1568

[24] Kristensen J, Zhang J, Chorkendorff I, Ulstrup J, Ooi BL. Assembled monolayers of Mo3S44+ clusters on well-defined surfaces. Dalt Trans. 2006;**33**:3985

[25] Kadirov MK, Knyazeva IR, Nizameev IR, Safiullin RA, Matveeva VI, Kholin KV, et al. Oxygen reduction reaction catalyzed by nickel complexes based on thiophosphorylated calix[4] resorcinols and immobilized in the membrane electrode assembly of fuel cells. Dalt Trans. 2016;**45**(41): 16157-16161

[26] Genorio B, Subbaraman R, Strmcnik D, Tripkovic D, Stamenkovic VR, Markovic NM. Tailoring the selectivity and stability of chemically modified platinum Nanocatalysts to design highly durable anodes for PEM fuel cells. Angew Chemie Int Ed. 2011;**50**(24):5468-5472

[27] An Q, Li G, Tao C, Li Y, Wu Y, Zhang W. A general and efficient method to form self-assembled cucurbit[n]uril monolayers on gold surfaces. Chemical Communications. 2008;**17**:1989

[28] Kaifer A, Gomez-Kaifer M. Supramolecular Electrochemistry. In:

Kaifer A, Gomez-Kaifer M, editors. Supramolecular Electrochemistry. Weinheim, Germany: Wiley-VCH Verlag GmbH; 1999. pp. 1-241

[29] Kolivoška V, Mohos M, Pobelov IV, Rohrbach S, Yoshida K, Hong WJ, et al. Electrochemical control of a non-covalent binding between ferrocene and beta-cyclodextrin. Chemical Communications. 2014;**50**(79): 11757-11759

[30] Crini G, Fourmentin S, Fenyvesi É, Torri G, Fourmentin M, Morin-Crini N. Cyclodextrins, from molecules to applications. Environmental Chemistry Letters. 2018;**16**(4):1361-1375

[31] Sévery L, Szczerbiński J, Taskin M, Tuncay I, Brandalise Nunes F, Cignarella C, et al. Immobilization of molecular catalysts on electrode surfaces using host–guest interactions. Nature Chemistry. 2021;**13**(6):523-529

[32] Hashidzume A, Yamaguchi H, Akira H. Cyclodextrin-based molecular machines. TripleC. 2014;**11**(1):13-35

[33] Del Valle EMM. Cyclodextrins and their uses: A review. Process Biochemistry. 2004;**39**(9):1033-1046

[34] Komatsu T, Hayakawa S, Tsuchida E, Nishide H. Meso-Tetrakis[o-(N-methyl) pyridinium]porphyrin ensembles with axially coordinated cyclodextrin-penetrating phenethylimidazole: Reversible dioxygen-binding in aqueous DMF solution. Chemical Communications. 2003;**1**:50-51

[35] D'Souza F, Hsieh Y-Y, Wickman H, Kutner W. New sensor for dissolved dioxygen: A gold electrode modified with a condensation polymer film of β-cyclodextrin hosting cobalt tetraphenylporphyrin. Chemical Communications. 1997;**13**:1191-1192

[36] Damos FS, Luz RCS, Tanaka AA, Kubota LT. Dissolved oxygen amperometric sensor based on layer-by-layer assembly using host–guest supramolecular interactions. Analytica Chimica Acta. 2010;**664**(2):144-150

[37] D'Souza F, Hsieh YY, Wickman H, Kutner W. β-Cyclodextrin and Carboxymethylated β-Cyclodextrin polymer film modified electrodes, hosting cobalt porphyrins, as sensors for Electrocatalytic determination of oxygen dissolved in solution. Electroanalysis. 1997;**9**(14):1093-1101

Section 3

Nanoparticles Self-Assembly and Their Applications

Chapter 5

Increasing Trend of Silver Nanoparticles as Antibacterial and Anticancer Agent

Attique Ur Rehman Khan, Muhammad Adnan, Shaheen Begum, Ruqia Nazir and Sakina Mussarat

Abstract

Silver nanoparticles (AgNPs) synthesis from plants that already have been reported for medicinal purposes demonstrated better efficacy for curing diseases. Recently, a number of researches have been reported where AgNPs act as promising antibacterial and anticancer agent. Biosynthesized silver nanoparticles (AgNPs) are a type of environmentally friendly, cost-effective, and biocompatible substance that has gotten a lot of attention in treatment of cancer and inhibition of pathogenic microbes. In this chapter, a comprehensive report on the recent development of AgNPs as nanomedicine synthesized from plant extracts. The role and mechanism of AgNPs as antibacterial and anticancer agent was reported that leads towards development of targeted nannomedicines to treat infectious diseases and world most challenging disease like cancer. Reported literature give imminence importance of AgNPs and demonstrated more potency to treat cancer and bacterial infections.

Keywords: silver nanoparticles, biolabeling, conjugation, phagocytosis, polydispersity

1. Introduction

Silver has been used widely from ancient times as it is a noble metal. Hippocrates advocated for the use of silver in the treatment of sickness and for healing purposes [1]. Silver is found abundantly in nature with multiple biological and biochemical properties making silver most suitable candidate for the biomedical applications, can be used as an antiseptic, part of medicines, antimicrobial efficacy, pharmaceutical industry, Food preservation, cosmetics, biolabelling, and optical properties. AgCl and AgNO3, ionic forms of silver, caused cardiac alterations in rats, such as left ventricular hypertrophy, hypersensitivity, and inhibition of normal fibroblast function [2]. Silver nanoparticles (AgNPs) are comparatively safe and more effective in medical treatment to silver ions [3].

Recently, nanotechnology has played a critical role in biomedical, diagnosis, treatments, the industrial sector, scientific purpose, and environmental protection [4]. Nanomaterials have a size range of 1–100 nm, or particles with at least one dimension smaller than 100 nm [4]. Due to unique physicochemical and biological characteristics, such as large surface area to volume ratio, excellent surface plasmon resonance, conjugation with various ligands to obtain desired property, inhibition against microbes, potent toxicity towards cancer cells, catalytic operations, silver nanoparticles are one of the most widely studied metal nanoparticles for a variety of scientific purposes. Due to very small size they penetrate the blood capillaries and tissues and become more effective in cancer treatment. Moreover they carry the multiple drugs on their large surface area and have capability to modify and combine chemically. Antimicrobial and anticancer activities of green synthesized AgNPs is due to phytoconstituents attached on their surface [5]. Several research studies have been conducted a green method to synthesize a range of metallic nanoparticles in concern with growing worldwide burden of cancer that showed potential anticancer effect against a range of cancer cell types [6]. Unicellular or multicellular living organisms are typically with 10 μm, so AgNPs in small size (1–100 nm) can interact with cell wall of bacterial, viral, and fungal pathogens and their active nano-complexes can penetrate and break the external capsule. The permeability of the plasma membrane to small-sized AgNPs permits them to accumulate in cell compartments. Phagocytosis, endocytosis, or micropinocytosis is the uptake mechanisms of nanoparticles in eukaryotic cells [7].

The rising applications of AgNPs in field of oncology and microbiology, present chapter emphasizes the significant antibacterial and anticancer properties of AgNPs synthesized by the green approach, recent developments and finding new perspectives in nanomedicine. In comparison with other methods, Ram Prasad's methods have been shown to be better due to their slow kinetics and ability to manipulate crystal growth and stabilization in a better way. The biogenetic synthesis uses plant extracts in aqueous form to create noble nanoparticles, as the extracts contain more reducing agents than plants. The availability of silver nanoparticles and their various metabolites makes plant-mediated silver nanoparticle synthesis a preferable method [8]. There are several phyto-constituents that are believed to reduce silver ions, including tannins, terpenoids, flavonoids, ketones, aldehydes, amides, and carboxylic acids. Plant extracts (chemical composition, amount, conjugation method) and nanoparticles (type, size, shape, polydispersity, etc.) play an influential role in the properties of a bioconjugate method [9].

In addition to being expensive to manufacture, the silver ion method has not been demonstrated to be clinically effective in randomized controlled trials and cannot be used with oxidizing solutions such as hypochlorite or H_2O_2 [10]. There are several drawbacks to the generation of silver nanoparticles (AgNPs) using a tube furnace, including the fact that it occupies a large space, consumes a lot of energy, raises the temperature in the surrounding environment, and requires a lot of time to achieve thermal stability. To achieve a stable operating temperature, a tube furnace typically requires several kilowatts and several time of preheating [11]. The polysaccride method is very temperature sensitive because the binding between the silver nano particles is very weak. If the temperature is increased slightly then the reversible reaction is started and the separation of the silver nano particles is started so the nano particles are unstable [12].

2. Applications and importance of silver nanoparticles

Nanoparticles and nano-composites synthesized from plants containing noble metals, silver nanoparticles are widely used metal due to incredible potential and significant usage. The diverse chemical and physical nature of AgNPs suggests

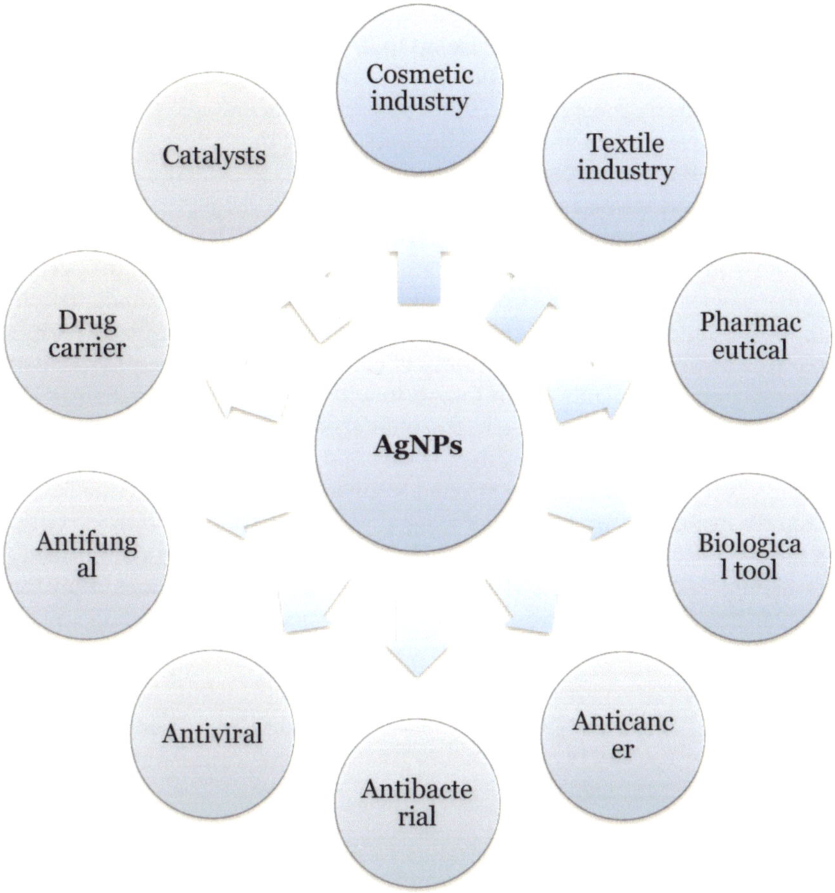

Figure 1.
Applications of AgNPs.

potential uses in the environment and for the well being of human life, promoting one health program for example cover the field of agriculture, food industry, medicine, and for the better human health (**Figure 1**) [13–15]. In the treatment of cancer cells, AgNPs are used as therapeutic agents due to cellular oxidative and apoptotic potential [15, 16]. AgNPs offer new uses due to their size-dependent actions and capacity to form various complexes with natural or synthetic molecules [17–19].

AgNPs are the most studied zero-dimensional nanoparticles for their remarkable and unparalleled uses in pharmaceutical science, infectious problems, wound care, antimicrobial, food packaging, and the cosmetic sector [20]. In recent years, biosynthesized AgNPs have shown potent larvicidal, bactericidal, fungicidal, antioxidant, antiviral, antidiabetic, and anticancer activities [21]. There are approximately 383 commercialized nano silver-based products on the market worldwide, accounting for 24% of all nano products [22].

2.1 Silver nanoparticles as anticancer agent

Cancer is one of the most challenging diseases to treat, and defined as the uncontrollable division of altered cells. It is the leading cause of mortality and about 70%

deaths in middle and low income countries and 68% population suffer due to cancer [23]. Globally cancer burden will be rise up to 27 million by 2040 [24]. Cancers most commonly diagnosed in the human population include lung, thyroid, cervical, liver, stomach, brain tumors, prostate, uterine, and breast cancers [19]. The most predominantly prevalent cancer are breast and prostate cancer that effect women and men, respectively. The field of cancer nanobiotechnology has provided new direction to detect, diagnose, and treat cancer [25]. AgNPs produced through green method with phytochemical covering give them more efficacy than AgNPs produced through chemical method. The ability to combine AgNPs inherent anticancer property with the pharmacological anticancer effects could be the key to treating malignancies that have stopped responding to chemotherapy or radiotherapy. Metal-based AgNPs have been found to be pro-oxidative in a variety of cancer cell types. The phytoconstituents berberine isolated from plants in combination with AgNPs showed synergistic anticancer activity [26]. Several studies in the published literature have looked into the methods by which AgNPs exhibit anticancer action. Among studied cancer cell lines most of the silver nanoparticles are studied against breast cancer cell line MCF-7. The size of AgNPs evaluated for anticancer ranged from 5 to 100 nm; with varying shapes such as spherical, cubical and hexagonal. The IC50 values of green synthesized AgNPs extracts against studied cell lines ranged from 6 to 1200 μg/mL. Some important studies regarding *in vitro* evaluation of AgNPs on cancerous cell lines represented in **Table 1**. AgNPs synthesized from *Mentha species* possess inhibitory effect against

Plant used	Part used	Size (nm) and shape of AgNPs	Cancer cell line	IC_{50} value (μg ml − 1)	Reference
Avicennia marina	Leaves	10 (spherical)	A549 lung cancer cells	15	[27]
Litchi chinensis	Leaves	59 (spherical)	MCF-7	40	[28]
Fagonia indica	Whole plant	10–60 (spherical)	MCF-7	12.3	
Ganoderma neo-japonicum	Fruit	5–8 (spherical)	MDA-MB-231	6	[29]
Putranjiva roxburgi	leaves	8 (spherical)	MDA-MB-231, HCT-116 and PANC-1	0.54–0.00025	[30]
Jasminum officinale	Rhizome	9.2 (spherical)	MCF-7, Bladder (5637)	9.3–1,13.0	[31]
Noctiluca scintillans	callus	4.2 (spherical)	MDA-MB-231	50	[32]
Euprenolepisprocera	Leaves	60 (spherical)	MCF-7	9.63	[19]
Nostoclinckia	Whole	9.39–25.89 (spherical)	MCF-7	27.79	[33]
Solanumtrilobatum	Seed coat	41.90 (spherical)	MCF-7	30	[34]
Elephantopusscaber	Peel	59 (spherical and polygonal)	Colo-259	17.4	[35]
Zingiberofficinale	seed	20–51 (spherical)	HT-29	150.8	[36]
Chlorophytumborivilianum	peel	52 (spherical)	HT-29	7	[37]
Oleachrysophylla, Lavandula dentate	Aerial parts	328.6–284.5 (spherical)	HCT116	99.35	[38]

Plant used	Part used	Size (nm) and shape of AgNPs	Cancer cell line	IC_{50} value (μg ml − 1)	Reference
Mentha arvensis	Leaves	100 (spherical)	HCT116	1.7	[39]
Zanthoxylumrhesta	Leaves	10–68 (spherical)	A549	65.17	[40]
Punicagranatum	leaves	6–45 (spherical)	A549	5	[41]
Derris trifoliate	leaves	16.92 (spherical)	A549	86.23	[42]
Dimocarpuslonganlour	Peel	8–22 (spherical)	H1299	5.33	[43]
Neptadeflersiana	Aerial parts	33 (cubical)	HeLa	23	[44]
Detariummicrocarpum	Leaf	81 (cubical)	HeLa, PANC-1	31.5–1.84	[45]
Ginkgo biloba	leaf	40 (spherical)	HeLa, SiHa	Dose dependent	[46]
Rhizophoraapiculata	leaf	>100 (spherical)	HEK-293, HeLa	0.062–1.98	[47]
Punicagranatum	leaf	46.1 (spherical)	HeLa	100	[41]
Allium sativum	leaf	100–800 (spherical)	HePG2	31.25	[48]
Biergavuaerecta	leaf	15.9 (spherical)	PA-1	25	[46]
Alternantherasessilis	Whole plant	30–50 (spherical)	PC3	6.8	[49]
Dimocarpuslongan	Peel	9–32 (cubical)	PC3	10	[43]
Perillafrutescens	Leaf	25.71 (spherical and cubical)	COLO205, LNCaP	39.28–24.33	[50]
Salvia miltiorrhiza	leaf	100 (spherical and hexagonal)	LNCaP	50	[51]
Zingiber officinale	Leaf	18.93 (spherical)	AsPC-1PANC-1	312–1295	[36]
Punicagranatum	leaf	35–69 (spherical)	HePG2	70	[41]
Elephantopusscaber	leaf	59 (spherical)	MCF-7, A-549 and SCC-40	10	[35]
Tamarindus indica	Fruit shell	20–30 (spherical)	MCF-7	120	[24]
Chaetomorphalinum		Smaller size (spherical)	HCT-116	48.84	[52]
Andrographis paniculata	Stem	Small size u-shaped	Vero cells	31.25	[53]
Phyllanthus niruri	Leaf			125	
Tinospora cordifolia	Leaf			250	
Conocarpus Lancifolius	Fruits	21 to 173 nm	MDA MB-231	16.8 μg/ml.	[54]
Tridax procumbens	Leaf	11.1–45.4	A459	42.70	[55]
Sambucus ebulus	Leaf	35–50	AGS and MCF-7	240	[56]
Parthenium hysterophorus	Leaf	20 spherical	HepG2	50	[57]
Cleome viscosa	leaf	20–50	A549 and PA1	28 and 30 μg/mL	[58]
Bee pollens	Leaf	44	MCF-7	90	[59]
chitosan,	Leaf	23	MDA-MB-231	4.6	[60]

Plant used	Part used	Size (nm) and shape of AgNPs	Cancer cell line	IC_{50} value ($\mu g\ ml-1$)	Reference
Mimusops elengi	Fruit	43 spherical shape	HT-29 and MCF7	155 and 179	[61]
Gloriosa superba	Stem	7nm–14 spherical	A549	46.54	[62]
Luffa acutangula,	Leaves	8 spherical	MCF-7	90	[63]
			MDA-MB-231	65	
			U87	80	
			DBTRG,	90	
(Pistacia terebinthus)	Leaves	32 spherical shape	MCF-7	25	[58]
Hypericum Perforatum	Leaves	100	HeLa	7.71	[64]
			Hep G2	12.44	
Alternanthera sessilis	Aerial part	10–30 nm/ spherical	MCF-7	3.04	[65]
Alternanthera tenella	Leaf	48 nm/–	MCF-7	42.5	[55]
Andrographis echioides	Leaf	68.06 nm/cubic, pentagonal, hexagona	MCF-7	31.5	[65]
Achillea biebersteinii	Flower	12 nm/spherical, pentagonal	MCF-7	20 μg/mL	[66]
Azadirachta indica	Leaf	40 nm/spherical	MCF-7	10	[59]
oriandrum sativum	Leaf	37 nm/spherical, rod, triangular, hexagonal	MCF-7	30.5	[67]
Citrullus colocynthis	Leaf	7.39 nm/spherical	MCF-7	2.4 μg/mL	[68]
Dendrophthoe falcata	Leaf	5–45 nm/spherical	MCF-7	7	[57]
Erythrina indica	Root	20–118 nm/ spherical	MCF-7	23.89	[60]
Melia dubia	Leaf	7.3 nm/irregular	MCF-7	31.2	[69]
Olax scandens		30–60 nm/ spherical	MCF-7	30	[70]
Piper longum	Root	46 nm/spherical	MCF-7	67	[71]
Quercus (genus)	Fruit hull	46 spherical	MCF-7	50	[34]
Rheum emodi	Root	27.5 nm/spherical	MCF-7	28	[47]
Sesbania grandiflora	Leaf	22 nm/spherical	MCF-7	20	[72]
Solanum trilobatum	fruit	41.90 nm/ spherical, polygonal	MCF-7	30	[46]
Syzygium cumini	Fruit	40 nm/spherical	MCF-7	10	[73]
Syzygium aromaticum	Fruit	5–20 nm/spherical	MCF-7	70	[74]
Tabernae montana divaricate	Leaf	Mean 22.85 nm/ spherica	MCF-7	20	[75]

Plant used	Part used	Size (nm) and shape of AgNPs	Cancer cell line	IC_{50} value (μg ml − 1)	Reference
Taxus baccata	Needles	Mean 75.1 nm/ spherical	MCF-7	0.25	[76]
Ulva lactuca	Whole	56 nm/spherica	MCF-7	37	[71]
Butea monosperma	LEAF	20–80 nm/ spherical	HNGC2	67	
Azadirachta indica	Leaf	2–18 nm/ triangular, hexagonal	shia	4.1	[77]
Melia azedarach	Leaf	78 nm/cubical, spherical	HeLa	300	[78]
Citrullus colocynthis	Leaf	16.57 nm/Spherical	HCT-116	30	[79]
Gymnema sylvestre	Leaf	Spherical	HT29	85	[80]

Key: A-549; H1299; lung cancer, MCF-7, MDA-MB-231; breast cancer cell lines, SCC-40; oral cancer, HCT-116, HT-29; colon cancer, PANC-1; pancreatic cancer, Bladder (5637); bladder cancer, HeLa & SiHa; cervical cancer, HEK-293; human embryonic kidney cells, HepG2; liver cancer, PA-1; ovarian teratocarcinoma cell, PC3; prostate cancer, and LNCaP; prostate adenocarcinoma.

Table 1.
Anticancer activities of silver nanoparticles (AgNPs) synthesized from plants.

HCT116 colon cancer cells in human by inhibiting the cell division and reducing G1 phase [18]. In another study green synthesized AgNPs by using plant extract showed potent cytotoxicity against lung cancer [23]. AgNPs synthesized from fruit of *Tamarindus indica* and *Nepeta deflersiana* resulted into apoptosis and cytotoxicity for human breast cancer and cervical cancer, respectively. A dose dependent anticancer effect was observed may be induced due to oxidative stress that leads to mitochondrial and DNA impairment [24]. Cell lines from liver, gastric, and prostate cancer showed cytotoxic effects against AgNPs from lotus plants [26]. AgNPs prepared from *Crataegus microphylla* (fruit) and *Gossypium hirsutum* (leaf) showed considerable distortion of gastric adenocarcinoma cells [81]. It has been reported that AgNPs target the lung adenocarcinoma cells breaking DNA helix, chromosomal instability, and damage the mitochondria of cancerous cells [81]. When AgNPs applied to MCF-7, it changes the morphological parameter modifications, inhibition of cell growth and significant loss of plasma membrane integrity.

Although AgNPs of large size >100 nm can be more effective but small size <10 nm penetrate the cell, get localized inside the nucleus easily and can induce cytotoxicity at greater level as reported by Avalos et al. that smaller size nanoparticles exhibit more cytotoxicity than larger size in MTT assay and lactate dehydrogenase assays [82]. The mechanism involved behind inducing cytotoxicity is (i) interruption in cellular respiration and DNA replication due to uptake of free silver ions (ii) production of free silver radicals and reactive oxygen species (ROS) (iii) damage to cell membrane [83]. AgNPs induce ROS production and reduce glutathione (SGH), nuclear factor kB (NF-kB) and tumor necrosis factor-alpha (TNF-1) levels within cells). Increasing levels of superoxide radicals disrupt the mitochondrial signal transduction pathway, resulting in apoptosis [84]. The increase level of reactive oxygen species and decrease glutathione elicit damage to different components of cell such as breaking of DNA, peroxidation of lipid membrane and protein carbonylation. Apoptosis occurs when caspases 3 and 9 are activated as a result of changing mitochondrial membrane potential. After that, it activates c-Jun NH2terminal kinase (JNK), which causes DNA breaks to cause cell cycle arrest and the creation of apoptotic bodies [85]. AgNPs prepared from plants increase the sub-G1 phases of cell cycle and exhibit potent cytotoxicity. Chang et al. demonstrated link between sub-G1 arrests in cancer cells treated with curcumin showed more apoptosis suggested that AgNPs induced apoptosis in cancerous cells by prolonged sub-G1 phase [86]. This implies that the enhanced sub-G1 arrest of cancerous cells, which is connected to the induction of apoptosis, may be resulting in the death of cancer cells due to AgNPs application. In addition, green synthesized AgNPs prevented the formation of new cells induced by vascular endothelial growth factor (VEGF). After penetrating into the cell, AgNPs inhibited VEGF and through Src-dependent pathway the vascular permeability 1 L-1βinduced occured. [87]. Due to this anti-angiogenic efficacy AgNPs recommended as a new gateway of treatment for cancer. Another mechanism suggested for the anticancer potential of AgNPs is autophagy-induced cell breakdown, which results in cell death. Additionally, because autophagolysosomes accumulate in cancer cells and are more prevalent there, greenly produced AgNPs encourage autophagy, which ultimately results in cell death [30].

2.2 Silver nanoparticles as antibacterial agent

Silver nanoparticles have antibacterial properties and they auspiciously appear to be more potent and efficient antimicrobial agents than other nanomaterials from noble metals, due to their unique properties such as a large surface to volume ratio, toxicity, interaction with phosphorus and sulfur compounds in the cell [88]. These characteristics make them excellent agents for treating a variety of microbial infectious complaints, as well as for overcoming microbial resistance to conventional medicines, whether used in single or in combination with other therapeutic formulations [89]. The synergistic action of nano-silver and a broad variety of phytoconstituents exhibit wide range of antibacterial qualities, as silver nanoparticles are easily manufactured from plant extracts with extraordinary stability and eco-friendly approach. According to a report antimicrobial agent containing silver ions can damage the external membrane of targeted cell by reacting with proteins (thiol group) resulted in inactivation of bacterial enzymes. Silver reduces DNA replication and uncouples electron transport from oxidative phosphorylation when applied. As a result it interferes with membrane permeability and inhibits the respiratory chain enzymes and kills the microbes at very low concentration [90, 91]. AgNPs have suppressed the growth of bacteria at the minimum inhibitory concentration (MIC) for example; *Cestrum nocturnum*at at 0.25 μg/ml concentration showed 36 mm zone of inhibition against *Citrobecter* which support the above statement. At a dosage of 10 μg/ml, *B. vulgaris* and *B. nigra* demonstrated substantial antibacterial activity against *S. aureus* (93 mm) while *Ceratonia siliqua* leaves showed 8 mm inhibitory zone against *E. coli* (**Table 1**). *Ocimum sanctum* at 5 μg/ml concentration showed 11 mm zone of inhibition against *E. coli*. When tested at minimum inhibitory concentrations, AgNPs showed excellent permeability through bacterial cell walls and plasma membranes. AgNPs interacting with plasma membranes and releasing Ag + ions into cell cytoplasm, Thus, respiratory mechanisms and ion exchange processes were disrupted in bacterial membranes and mesosomes, and the obstruction of sulfur-containing protein synthesis on ribosomes [92]. When biologically produced nanoparticles and

$AgNO_3$ solutions were combined, the cytotoxic action was enhanced [93]. The addition of Ag + ions to the culture media reduced the production of biofilms by bacteria during growth. In biological experiments, AgNPs were found to have anti-biofilm formation properties against Gram positive (*Enterococcus faecalis* and *S. aureus*) and Gram negative (*Shigella sonnei* and *Pseudomonas aeruginosa*) [94]. All of these mechanisms of action show that AgNPs have antibacterial capabilities and that they can be used as anti-pathogenic drugs to reduce microorganism proliferation (**Table 2**).

Plant used	Concentration (μg/ml)	Bacteria	ZI (mm)	Reference
Coptis chinensis	12.50	*E. coli*	12	[95]
Cestrum nocturnum	0.25	*Citrobacter*	36	[94]
	1	*S. typhi*	28	
	2	*E. faecalis*	15	
	4	*E. coli*	23	
	8	*P. vulgaris*	26	
	16	*V. cholerae*	41	
B. vulgaris	10	*P. aeruginosa*	77.57	[96]
		E. coli	89.21	
		S. aureus	93.64	
B. nigar	10	*P. aeruginosa*	73.83	
		E. coli	83.31	
		S. aureus	93.12	
C. burspastoris	15.50	*P. aeruginosa*	92.62	
		E. coli	80.76	
		S. aureus	96.03	
Ceratonia siliqua	10	*E. coli*	8	[97]
Helictere sisora	12.5	*E. coli*	2	[98]
	100	*V. cholerae*	6	
Ocimum sanctum	5	*E. coli*	11	[99]
		S. aureus	10	
Acalypha indica	10	*E. coli*		[100]
Citrus aurantiifolia		*E. coli*	7	[101]
Citrus sinensis			8	
Citrus limetta			6	
Citrus aurantiifolia		*K. pneumoniae*	6	
Citrus sinensis			8	
Citrus limetta			5	
Citrus aurantiifolia		*S. aureus*	5	
Citrus sinensis			5	
Citrus limetta			4	

Plant used	Concentration (μg/ml)	Bacteria	ZI (mm)	Reference
Citrus aurantiifolia		*S. typhimurium*	6	
Citrus sinensis			6	
Citrus limetta			4	
Zingiber officinale	100	*Staphylococcus spp*	6.5	[102]
Coffea arabica	0.05	*E. coli*	2.3	[103]
	0.1		3.1	
	0.05	*S. aureus*	2.1	
	0.1		2.7	
Chlorophytum borivilianum	15	*S. aureus*	10	[37]
	30		17	
	60		29	
	15	*P. aeruginosa*	9	
	30		11	
	60		14	
Ficus sycomorus	50	*E. coli*	9	[104]
		S. aureus	11	
		P. aeruginosa	11	
		K. pneumoniae	18	
		S. typhi	30	
		S. flexneri	16	
Zataria multiflora	20	*S. aureus*		[105]
		P. aeruginosa		
Malva verticillata	100	*A. hydrophila n*	12.44	[106]
		A. salmonicida	28.64	
Camilla sinensis	50	*E. coli*	12.5	[107]
Rhizophora apiculata	21	*B. subtilis*	11	[108]
		E. coli	14	
		K. pneumoniae	14	
		P. vulgaris	14	
		P. aeruginosa	12	
		S. typhi	14	
Dryopteris crassirhizoma	100	*B. cereus*	2	[109]
	150	*P. aeruginosa*	3	
Acacia leucophloea	25	*S. aureus*	15	[110]
	50	*B. cereus*	17.50	
	75	*S. flexneri*	17	
Olea europaea	7	*S. aureus*	2.7	[111]

Plant used	Concentration (μg/ml)	Bacteria	ZI (mm)	Reference
Conocarpus Lancifolius	2.5	*s. aureus*	4	[54]
	5		7.5	
	10		11	
	20		14	
	50		22	
	2.5	*s. pneumoniae*	3.8	
	5		7	
	10		10	
	20		11	
	50		19	
Tridax procumbens	20	*E. coli*	11	[55]
		Shigella. ssp	15	
		Pseudomonas aeruginosa	20.66	
		Pseudomonas aeruginosa	15.33	
		Candida tropicalis	20	
Cleome viscosa	10	*s. aureus*	11	[101]
	20		13	
	30		14	
	40		17	
	10	*B. subtilis*	10	
	20		12	
	30		13	
	40		14	
	10	*E. coli*	10	
	20		13	
	30		15	
	40		16	
Bee pollen	100 μg/mL	*B. subtilis*	18	[59]
		P. aeruginosa	18	
		S. aureus	17	
		E. coli	11	
Parthenium hysterophorus	60	*E. coli*	17	[60]
		P. aeruginosa	18	
		B. subtilis	12	
		E. feacali	11	
		S. aureus	15	

Plant used	Concentration (μg/ml)	Bacteria	ZI (mm)	Reference
Gloriosa superba	40	*Enterococcus faecalis*	29	[62]
		Bacillus subtilis	24	
		Staphylococcus aureus	23	
Luffa acutangula	5	*B. subtilis*	7.2	[63]
		S. aureus	7.9	
		E. coli	7.4	
P. americana	25	*P. vermicola*	24	[102]
		A. caviae	17	
	50	*E. coli*	10	[103]
		Bacillus subtilis	8	
Taxus baccata Linn	25	*Shigella dysenteriae*	10	[77]
		E. coli	12	
		Salmonella typhi	08	
Gardenia thailandica	50	*S. aureus*	12	[103]

Table 2.
Antibacterial activity of silver nanoparticles (AgNPs) from plants.

2.3 Antifungal activity of silver Nano particles

Drug resistance by pathogenic fungi has been continuously increasing, so it is necessary to develop new antifungal agent. The antifungal agent was present in the form of the chemically, physically and, biologically. The green plants which caring affective metabolites and particles which use against the fungus disease. There are the many nano particles which use against the fungi but Silver nano particles have the drastic affect against the many disease which is caused by the fungi [25]. In many study reported that the AgNPs as antifungal agent in treating fungal infectious diseases [112]. This disease badly affected the human and the plants as well. Silver nanoparticle are very effective against the four pathogens *R. solani, F. oxysporum, S. sclerotiorum,* and *S. rolfsii* which caused the disease in the vegetable and horticulture. Silver nanoparticles activity was checked at the four different concentration against these fungus species [113]. Silver nano particles was seen the most activity against the candida albican species include *C. tropicalis*, *C. glabrata*, *C. parapsilosis. C. glabrata* at different concentration (0.01 μg −300 μg) with different zone of inhibitions (05–70) [114] (**Table 3**).

Plant name	Concentration μg	Fungus		Reference
Garcinia kola pulp	75	*Candida tropicalis*	13	[115]
		Fusarium oxysperium	15	
Taxus baccata Linn	90	*T. purpureogenus*	22	[77]
Juniperus procera	50	*C. albicans* ATCC885653	14.3	
	50	*C. neoformans* ATCC16620	9.80	

Plant name	Concentration µg	Fungus		Reference
red curran	30	*Fusarium oxysporum*	12	[116]
		Botrytis cinerea	26	
		Pestalotiopsis mangiferae	50	
E. tirucalli	923.4	*B. cinérea*	1.9	[117]
		R. stolonifera	3.5	
B. lanzan Spreng	50 ppm	*Rhizoctonia solani.*	47	[118]
	100 ppm		52	
Allium fistulosum	0.32 mg/mL	*Aspergillus niger*	08	[119]
	10 mg/mL		11	
	0.32 mg/mL	*Candida albicans*	07	
	10 mg/mL		10	
Cynara cardunculus	1.8 mg/mL	*C. albicans*	26.6	[120]
Glycosmis pentaphylla	11	*A. alternata*	15.5	[121]
		F. moniliforme	14.5	
		Colletotrichum lindemuthianum	9.5	
		Candida glabrat	12	
O. vulgare	5 µL	*Aspergillus flavus*	7.7	[122]
	10 µL		11	
	5 µL	*Fusarium moniliform*	7	
	10 µL		10	
	5 µL	*Candida albicans*	10	
	10 µL		18	
Borago officinalis	100	*Candida albicans*	7	[123]
Alhagi graecorum	0.01 mmol\ml	*C. albicans*	14	[124]
		C. glabrata	18	
		C. parapsilosis	22	
		C. tropicales	21	
		C. krusei	15	
	0.02 mmol\ml	*C. albicans*	16	
		C. glabrata	21	
		C. parapsilosis	27	
		C. tropicales	25	
		C. krusei	17	
Malva parviflora L	15 µg /mL	*F. solani*	81.1	[125]
		A. alternata	83	
		H. rostratum	88.6	
		F. oxysporum	80	

Plant name	Concentration µg	Fungus		Reference
Allium ampeloprasum	25 µg /mL	*C. albicans*	20.1	[126]
		C. glabrata	20.6	
		C. krusei	15.1	
		C. tropicalis	16.4	
		C. parapsilosis	18.4	
Melia azedarach	20 µg /mL	*Verticillium dahliae*	87	[127]
Teucrium polium L	50 µg /mL	*F. oxysporum*	46	[128]
	100 µg /mL		54	
	150 µg /mL		54	
plant essential oil	20 µg /mL	*Aspergillus niger*	11.33	[129]
	20 µg /mL	*Aspergillus flavus*	13.27	
		Candida albicans	9.87	
		Candida tropicalis	14.66	
		Candida kefyr	15.17	
maize	25	*C. albicans*	0.021	[130]
Maize	47 g	*Candida albicans*	62.5	[131]
Ferulago macrocarpa	250 µg/mL	*Candida albicans*	34	[132]
Lotus lalambensis	6.25	*Candida albicans*	10	[133]
	12.5		13	
	25		16	
	50		19	
Grass waste	2	*F. solani*	20	[134]
	5		38	
	10		60	

Table 3.
Antifungal activity of silver nanoparticles (AgNPs) from plants.

3. Conclusions

Due to the vast range of activities and unique physical and chemical characteristics, silver nanoparticles are currently the subject of in-depth research. AgNPs are effective anticancer agents because they affect the cell cycle, prevent the growth of cancer cells, cause oxidative stress, and promote apoptosis [135, 136]. They protect against bacterial infections and showed potent antibacterial effect at minute concentrations. Due to the weakened immunological resistance of cancer patients, such antimicrobial protection is preferred during chemo- and radiotherapy. Most of the literature for use of AgNPs as antibacterial and anticancer agent is quite reported recently in present century showing that nanomedicine has made many advances in ongoing years and still there need to explored this field [137–139]. In order to gain unique insights and improve silver NP characteristics, additional research on AgNPs needs to be done.

Future applications may involve certain contentious concerns, like dose for various tissues; side effects from therapy, tissue-specific biocompatibility, or microbial resistance to NPs. AgNPs have some actions that seem to be dual or even contradictory depending on the situation. Examples include anti- or pro-oxidative, biosensing or bioresisting activity depending on the type of cell or living organism. Before being added to cells, NPs must be thoroughly described and their physical and chemical characteristics must be understood. These characteristics are mostly the product of various AgNP synthesis techniques, and only nontoxic ones should be favored in bioassays involving living models.

Author contributions

Muhammad Adnan and Ruqia Nazir, Sakina Mussarat conceived the idea of chapter, helped in writing and provide useful suggestions. Sakina Mussarat and Attique ur Rehman Khan participated in writing of the manuscript, and performed all literature surveys, designed the figures and reviewed the literature. All authors were involved in revising the chapter content, read, and approved the final draft.

Conflict of interest

The authors declared no potential conflicts of interest.

Author details

Attique Ur Rehman Khan[1*†], Muhammad Adnan[2*†], Shaheen Begum[3], Ruqia Nazir[1] and Sakina Mussarat[2†]

1 Department of Chemistry, Kohat University of Science and Technology, Kohat, Pakistan

2 Department of Botany, Kohat University of Science and Technology, Kohat, Pakistan

3 Department of Environmental Sciences, Fatima Jinnah Women University, Rawalpindi, Pakistan

*Address all correspondence to: atteeqrehmankhan78@gmail.com and ghurzang@hotmail.com

† These authors contributed equally.

References

[1] Russell A, Hugo W. 7 antimicrobial activity and action of silver. Progress in Medicinal Chemistry. 1994;**31**:351-370

[2] Jamieson WE, Fradet GJ, Abel JG, Janusz MT, Lichtenstein SV, MacNab JS, et al. Seven-year results with the St Jude Medical Silzone mechanical prosthesis. The Journal of Thoracic and Cardiovascular Surgery. 2009;**137**(5): 1109-1115. e1102

[3] Ge L, Li Q, Wang M, Ouyang J, Li X, Xing MM. Nanosilver particles in medical applications: Synthesis, performance, and toxicity. International Journal of Nanomedicine. 2014;**9**:2399

[4] Farokhzad OC, Langer R. Impact of nanotechnology on drug delivery. ACS Nano. 2009;**3**(1):16-20

[5] Tadele KT, Abire TO, Feyisa TY. Green synthesized silver nanoparticles using plant extracts as promising prospect for cancer therapy: A review of recent findings. Journal of Nanomedicine. 2021;**4**:1040

[6] Jha M, Shimpi NG. Green synthesis of zero valent colloidal nanosilver targeting A549 lung cancer cell: in vitro cytotoxicity. Journal of Genetic Engineering and Biotechnology. 2018;**16**(1):115-124

[7] Skonieczna M, Hudy D. Biological Activity of Silver Nanoparticles and Their Applications in Anticancer Therapy. London, UK, London: IntechOpen; 2018

[8] Choi S, Kim KS, Yeon SH, Cha JH, Lee H, Kim CJ, et al. Fabrication of silver nanoparticles via self-regulated reduction by 1-(2-hydroxyethyl)-3-methylimidazolium tetrafluoroborate. Korean Journal of Chemical Engineering. 2007;**24**(5):856-859

[9] Park Y et al. A new paradigm shift for the green synthesis of antibacterial silver nanoparticles utilizing plant extracts. Toxicological Research. 2014;**30**(3): 169-178

[10] Li X, Xu H, Zhe-Sheng C, et al. Biosynthesis of nanoparticles by microorganisms and their applications. Journal of Nanomaterials. 2011;**11**:1-16

[11] Abou El-Nour KM et al. Synthesis and applications of silver nanoparticles. Arabian Journal of Chemistry. 2010;**3**(3): 135-140

[12] Jung J, Oh H, Noh H, Ji J, Kim S. Metal nanoparticle generation using a small ceramic heater with a local heating area. Journal of Aerosol Science. 2006; **37**(1):662-167

[13] Ahmed KBR, Nagy AM, Brown RP, Zhang Q, Malghan SG, Goering PL. Silver nanoparticles: Significance of physicochemical properties and assay interference on the interpretation of in vitro cytotoxicity studies. Toxicology In Vitro. 2017;**38**:179-192

[14] Bouwmeester H, Dekkers S, Noordam MY, Hagens WI, Bulder AS, De Heer C, et al. Review of health safety aspects of nanotechnologies in food production. Regulatory Toxicology and Pharmacology. 2009; **53**(1):52-62

[15] Aueviriyavit S, Phummiratch D, Maniratanachote R. Mechanistic study on the biological effects of silver and gold nanoparticles in Caco-2 cells–induction of the Nrf2/HO-1 pathway by high concentrations of silver nanoparticles. Toxicology Letters. 2014; **224**(1):73-83

[16] Rai M, Ingle AP, Birla S, Yadav A, Santos CAD. Strategic role of selected noble metal nanoparticles in medicine. Critical Reviews in Microbiology. 2016; **42**(5):696-719

[17] Mfouo-Tynga I, El-Hussein A, Abdel-Harith M, Abrahamse H. Photodynamic ability of silver nanoparticles in inducing cytotoxic effects in breast and lung cancer cell lines. International Journal of Nanomedicine. 2014;**9**(1):3771-3780

[18] El-Hussein A, Hamblin MR. ROS generation and DNA damage with photo-inactivation mediated by silver nanoparticles in lung cancer cell line. IET Nanobiotechnology. 2017;**11**(2):173-178

[19] Hudecki A, Gola J, Ghavami S, Skonieczna M, Markowski J, Likus W, et al. Structure and properties of slow-resorbing nanofibers obtained by (co-axial) electrospinning as tissue scaffolds in regenerative medicine. PeerJournal. 2017;**5**:e4125

[20] Deng H, McShan D, Zhang Y, Sinha SS, Arslan Z, Ray PC, et al. Mechanistic study of the synergistic antibacterial activity of combined silver nanoparticles and common antibiotics. Environmental Science & Technology. 2016;**50**(16):8840-8848

[21] Calderón-Jiménez B, Johnson ME, Montoro Bustos AR, Murphy KE, Winchester MR, Vega Baudrit JR. Silver nanoparticles: Technological advances, societal impacts, and metrological challenges. Frontiers in Chemistry. 2017; **5**:6

[22] Rafique M, Rafique MS, Kalsoom U, Afzal A, Butt SH, Usman A. Laser ablation synthesis of silver nanoparticles in water and dependence on laser nature. Optical and Quantum Electronics. 2019; **51**(6):1-11

[23] Javed B, Nadhman A, Razzaq A. One-pot phytosynthesis of nano-silver from Mentha longifolia L.: Their characterization and evaluation of photodynamic potential. Materials Research Express. 2020;**7**(5):055401

[24] Jabeen S, Qureshi R, Munazir M, Maqsood M, Munir M, Shah SH, et al. Application of green synthesized silver nanoparticles in cancer treatment-a critical review. Materials Research Express. 2021

[25] Kowsalya E, MosaChristas K, Balashanmugam P, Rani JC. Biocompatible silver nanoparticles/poly (vinyl alcohol) electrospun nanofibers for potential antimicrobial food packaging applications. Food Packaging and Shelf Life. 2019;**21**(7):100-379

[26] Sung H, Ferlay J, Siegel RL, Laversanne M, Soerjomataram I, Jemal A, et al. Global cancer statistics 2020: GLOBOCAN estimates of incidence and mortality worldwide for 36 cancers in 185 countries. CA: a Cancer Journal for Clinicians. 2021;**71**(3): 209-249

[27] Tian S, Saravanan K, Mothana RA, Ramachandran G, Rajivgandhi G, Manoharan N. Anti-cancer activity of biosynthesized silver nanoparticles using Avicennia marina against A549 lung cancer cells through ROS/mitochondrial damages. Saudi Journal of Biological Sciences. 2020;**27**(11):3018-3024

[28] Iqbal MJ, Ali S, Rashid U, Kamran M, Malik MF, Sughra K, et al. Biosynthesis of silver nanoparticles from leaf extract of Litchi chinensis and its dynamic biological impact on microbial cells and human cancer cell lines. Cellular and Molecular Biology. 2018;**64**(13):42-47

[29] Gurunathan S, Raman J, Abd Malek SN, John PA, Vikineswary S.

Green synthesis of silver nanoparticles using Ganoderma neo-japonicum Imazeki: A potential cytotoxic agent against breast cancer cells. International Journal of Nanomedicine. 2013;**8**:4399

[30] Balkrishna A, Sharma VK, Das SK, Mishra N, Bisht L, Joshi A, et al. Characterization and anti-cancerous effect of Putranjiva roxburghii seed extract mediated silver nanoparticles on human colon (HCT-116), pancreatic (PANC-1) and breast (MDA-MB 231) cancer cell lines: A comparative study. International Journal of Nanomedicine. 2020;**15**:573

[31] Elhawary S, Hala E-H, Mokhtar FA, Mansour Sobeh EM, Osman S, El-Raey M. Green synthesis of silver nanoparticles using extract of Jasminum officinal l. leaves and evaluation of cytotoxic activity towards bladder (5637) and breast cancer (MCF-7) cell lines. International Journal of Nanomedicine. 2020;**15**:9771

[32] Elgamouz A, Idriss H, Nassab C, Bihi A, Bajou K, Hasan K, et al. Green synthesis, characterization, antimicrobial, anti-cancer, and optimization of colorimetric sensing of hydrogen peroxide of algae extract capped silver nanoparticles. Nanomaterials. 2020;**10**(9):1861

[33] El-Naggar NE-A, Hussein MH, El-Sawah AA. Bio-fabrication of silver nanoparticles by phycocyanin, characterization, in vitro anticancer activity against breast cancer cell line and in vivo cytotxicity. Scientific Reports. 2017;**7**(1):1-20

[34] Ramar M, Manikandan B, Marimuthu PN, Raman T, Mahalingam A, Subramanian P, et al. Synthesis of silver nanoparticles using Solanum trilobatum fruits extract and its antibacterial, cytotoxic activity against human breast cancer cell line MCF 7. Spectrochimica Acta Part A: Molecular and Biomolecular Spectroscopy. 2015;**140**:223-228

[35] Shinde AS, Mendhulkar V. Antiproliferative activity of Elephantopus scaber mediated silver nanoparticles against MCF-7, A-549, SCC-40 and COLO-205 human cancer cell lines. Asian Journal of Pharmaceutical and Clinical Research. 2020;**13**(2):163-167

[36] Venkatadri B, Shanparvish E, Rameshkumar M, Arasu MV, Al-Dhabi NA, Ponnusamy VK, et al. Green synthesis of silver nanoparticles using aqueous rhizome extract of Zingiber officinale and Curcuma longa: In-vitro anti-cancer potential on human colon carcinoma HT-29 cells. Saudi Journal of Biological Sciences. 2020;**27**(11): 2980-2986

[37] Huang F, Long Y, Liang Q, Purushotham B, Swamy MK, Duan Y. Safed Musli (Chlorophytum borivilianum L.) callus-mediated biosynthesis of silver nanoparticles and evaluation of their antimicrobial activity and cytotoxicity against human colon cancer cells. Journal of Nanomaterials. 2019;**2019**

[38] Sufyani A, Moslah N, Hussien NA, Hawsawi YM. Characterization and anticancer potential of silver nanoparticles biosynthesized from Olea chrysophylla and Lavandula dentata leaf extracts on HCT116 colon cancer cells. Journal of Nanomaterials. 2019;**2019**

[39] Banerjee PP, Bandyopadhyay A, Harsha SN, Policegoudra RS, Bhattacharya S, Karak N, et al. Mentha arvensis (Linn.)-mediated green silver nanoparticles trigger caspase 9-dependent cell death in MCF7 and MDA-

MB-231 cells. Breast Cancer: Targets and Therapy. 2017;**9**:265

[40] Nayaka S, Chakraborty B, Pallavi S, Bhat MP, Shashiraj K, Ghasti B. Synthesis of biogenic silver nanoparticles using Zanthoxylum rhetsa (Roxb.) DC seed coat extract as reducing agent and in-vitro assessment of anticancer effect on A549 lung cancer cell line. International Journal of Pharmaceutical Research. 2020;**12**(3):302-314

[41] Annu M, Ahmed S, Kaur G, Sharma P, Singh S, Ikram S. Evaluation of the antioxidant, antibacterial and anticancer (lung cancer cell line A549) activity of Punica granatum mediated silver nanoparticles. Toxicology Research. 2018;**7**(5):923-930

[42] Cyril N, George JB, Joseph L, Raghavamenon A, Vp S. Assessment of antioxidant, antibacterial and anti-proliferative (lung cancer cell line A549) activities of green synthesized silver nanoparticles from Derris trifoliata. Toxicology Research. 2019;**8**(2):297-308

[43] He Y, Du Z, Ma S, Cheng S, Jiang S, Liu Y, et al. Biosynthesis, antibacterial activity and anticancer effects against prostate cancer (PC-3) cells of silver nanoparticles using Dimocarpus Longan Lour. Peel extract. Nanoscale Research Letters. 2016;**11**(1):1-10

[44] Al-Sheddi ES, Farshori NN, Al-Oqail MM, Al-Massarani SM, Saquib Q, Wahab R, et al. Anticancer potential of green synthesized silver nanoparticles using extract of Nepeta deflersiana against human cervical cancer cells (HeLA). Bioinorganic Chemistry and Applications. 2018:1-12

[45] Adebayo IA, Arsad H, Gagman HA, Ismail NZ, Samian MR. Inhibitory effect of eco-friendly naturally synthesized silver nanoparticles from the leaf extract of medicinal Detarium microcarpum plant on pancreatic and cervical cancer cells. Asian Pacific journal of cancer prevention: APJCP. 2020;**21**(5):1247

[46] Xu Z, Feng Q, Wang M, Zhao H, Lin Y, Zhou S. Green biosynthesized silver nanoparticles with aqueous extracts of ginkgo biloba induce apoptosis via mitochondrial pathway in cervical cancer cells. Frontiers in Oncology. 2020;**10**:575415

[47] Liu X, Shan K, Shao X, Shi X, He Y, Liu Z, et al. Nanotoxic effects of silver nanoparticles on normal HEK-293 cells in comparison to cancerous HeLa cell line. International Journal of Nanomedicine. 2021;**16**:753

[48] Palle SR, Penchalaneni J, Lavudi K, Gaddam SA, Kotakadi VS, Challagundala VN. Green synthesis of silver nanoparticles by leaf extracts of boerhavia erecta and spectral characterization and their antimicrobial, antioxidant ad cytotoxic studies on ovarian cancer cell lines. Letters in Applied NanoBioScience. 2020;**9**: 1165-1176

[49] Firdhouse MJ, Lalitha P. Biosynthesis of silver nanoparticles using the extract of Alternanthera sessilis—Antiproliferative effect against prostate cancer cells. Cancer Nanotechnology. 2013;**4**(6):137-143

[50] Reddy N, Li H, Hou T, Bethu M, Ren Z, Zhang Z. Phytosynthesis of silver nanoparticles using Perilla frutescens leaf extract: Characterization and evaluation of antibacterial, antioxidant, and anticancer activities. International Journal of Nanomedicine. 2021;**16**:15

[51] Zhang K, Liu X, Samuel Ravi SOA, Ramachandran A, Aziz Ibrahim IA, Nassir AM, et al. Synthesis of silver nanoparticles (AgNPs) from leaf extract

of salvia miltiorrhiza and its anticancer potential in human prostate cancer LNCaP cell lines. Artificial Cells, Nanomedicine, and Biotechnology. 2019; **47**(1):2846-2854

[52] Acharya D, Satapathy S, Somu P, Parida UK, Mishra G. Apoptotic effect and anticancer activity of biosynthesized silver nanoparticles from marine algae Chaetomorpha linum extract against human colon cancer cell HCT-116. Biological Trace Element Research. 2021; **199**(5):1812-1822

[53] Sharma D et al. Antimicrobial and cytotoxic potential of silver nanoparticles synthesized using Rheum emodi roots extract. New Frontiers in Chemistry. 2015;**24**(2):121

[54] Oves M et al. Green synthesis of silver nanoparticles by Conocarpus Lancifolius plant extract and their antimicrobial and anticancer activities. Saudi Journal of Biological Sciences. 2022;**29**(1):460-471

[55] Pungle R et al. Green synthesis of silver nanoparticles using the Tridax procumbens plant extract and screening of its antimicrobial and anticancer activities. Oxidative Medicine and Cellular Longevity. 2022:1-14

[56] Hashemi Z et al. Sustainable green synthesis of silver nanoparticles using Sambucus ebulus phenolic extract (AgNPs@ SEE): Optimization and assessment of photocatalytic degradation of methyl orange and their in vitro antibacterial and anticancer activity. Arabian Journal of Chemistry. 2022; **15**(1):103525

[57] Ahsan A et al. Green synthesis of silver nanoparticles using Parthenium hysterophorus: Optimization, characterization and in vitro therapeutic evaluation. Molecules. 2020;**25**(15)

[58] Lakshmanan G et al. Plant-mediated synthesis of silver nanoparticles using fruit extract of Cleome viscosa L.: Assessment of their antibacterial and anticancer activity. Karbala International Journal of Modern Science. 2018;**4**(1): 61-68

[59] Al-Yousef HM et al. Pollen bee aqueous extract-based synthesis of silver nanoparticles and evaluation of their anti-cancer and anti-bacterial activities. PRO. 2020;**8**(5):52

[60] Venkatesan J et al. Antimicrobial and anticancer activities of porous chitosan-alginate biosynthesized silver nanoparticles. International Journal of Biological Macromolecules. 2017;**98**: 515-525

[61] Korkmaz N et al. Biogenic silver nanoparticles synthesized via Mimusops elengi fruit extract, a study on antibiofilm, antibacterial, and anticancer activities. Journal of Drug Delivery Science and Technology. 2020;**59**:101864

[62] Murugesan AK et al. Facile green synthesis and characterization of Gloriosa superba L. tuber extract-capped silver nanoparticles (GST-AgNPs) and its potential antibacterial and anticancer activities against A549 human cancer cells. Environmental Nanotechnology, Monitoring & Management. 2021;**15**: 100460

[63] Naghmachi M, Raissi A, Baziyar P, Homayoonfar F, Amirmahani F, Danaei M. Green synthesis of silver nanoparticles (AgNPs) by Pistacia terebinthus extract: Comprehensive evaluation of antimicrobial, antioxidant and anticancer effects. Biochemical and Biophysical Research Communications. 2022;**608**:163-169

[64] Firdhouse J, Lalitha P. Apoptotic efficacy of biogenic silver nanoparticles

on human breast cancer MCF-7 cell lines. Progress in Biomaterials. 2015;**4**:113

[65] Sathishkumar P et al. Phyto-synthesis of silver nanoparticles using Alternanthera tenella leaf extract: An effective inhibitor for the migration of human breast adenocarcinoma (MCF-7) cells. Bioprocess and Biosystems Engineering. 2016;**39**(4):651-659

[66] Sre PR et al. Antibacterial and cytotoxic effect of biologically synthesized silver nanoparticles using aqueous root extract of Erythrina indica lam. Spectrochimica Acta Part A: Molecular and Biomolecular Spectroscopy. 2015;**135**:1137-1144

[67] Kathiravan V et al. Synthesis of silver nanoparticles from Melia dubia leaf extract and their in vitro anticancer activity. Spectrochimica Acta Part A: Molecular and Biomolecular Spectroscopy. 2014;**130**:116-121

[68] Reddy N et al. Evaluation of antioxidant, antibacterial and cytotoxic effects of green synthesized silver nanoparticles by Piper longum fruit. Materials Science and Engineering: C. 2014;**34**:115-122

[69] Heydari R, Rashidipour M. Green synthesis of silver nanoparticles using extract of oak fruit hull (Jaft): Synthesis and in vitro cytotoxic effect on MCF-7 cells. International Journal of Breast Cancer. 2015:2015

[70] Jeyaraj M et al. Biogenic silver nanoparticles for cancer treatment: An experimental report. Colloids and Surfaces B: Biointerfaces. 2013;**106**:86-92

[71] Sathishkumar P et al. Anti-acne, anti-dandruff and anti-breast cancer efficacy of green synthesised silver nanoparticles using Coriandrum sativum leaf extract. Journal of Photochemistry and Photobiology B: Biology. 2016;**163**: 69-76

[72] Mittal AK et al. Comparative studies of anticancer and antimicrobial potential of bioinspired silver and silver-selenium nanoparticles. Journal of Materials NanoScience. 2016;**3**(2):22-27

[73] Venugopal K et al. Synthesis of silver nanoparticles (Ag NPs) for anticancer activities (MCF 7 breast and A549 lung cell lines) of the crude extract of Syzygium aromaticum. Journal of Photochemistry and Photobiology B: Biology. 2017;**167**:282-289

[74] Devaraj P et al. Synthesis and characterization of silver nanoparticles using Tabernae Montana divaricata and its cytotoxic activity against MCF7 cell line. International Journal of Pharmacy and Pharmaceutical Sciences. 2014;**6**(8): 86-90

[75] Kajani AA et al. Green synthesis of anisotropic silver nanoparticles with potent anticancer activity using Taxus baccata extract. RSC Advances. 2014; **4**(106):61394-61403

[76] Devi JS, Bhimba BV, Ratnam K. In vitro anticancer activity of silver nanoparticles synthesized using the extract of Gelidiella sp. International Journal of Pharmacy and Pharmaceutical Sciences. 2012;**4**(4):710-715

[77] Mishra A et al. Effect of biologically synthesized silver nanoparticles on human cancer cells. Science of Advanced Materials. 2012;**4**(12):1200-1206

[78] Sukirtha R et al. Cytotoxic effect of green synthesized silver nanoparticles using Melia azedarach against in vitro HeLa cell lines and lymphoma mice model. Process Biochemistry. 2012; **47**(2):273-279

[79] Shawkey AM et al. Green nanotechnology: Anticancer activity of silver nanoparticles using Citrullus colocynthis aqueous extracts. Advances in Life Science and Technology. 2013;**13**: 60-70

[80] Arunachalam KD et al. Potential anticancer properties of bioactive compounds of Gymnema sylvestre and its biofunctionalized silver nanoparticles. International Journal of Nanomedicine. 2015;**10**:31

[81] Kanipandian N, Li D, Kannan S. Induction of intrinsic apoptotic signaling pathway in A549 lung cancer cells using silver nanoparticles from Gossypium hirsutum and evaluation of in vivo toxicity. Biotechnology Reports. 2019;**23**: e00339

[82] Avalos A, Haza AI, Mateo D, Morales P. Interactions of manufactured silver nanoparticles of different sizes with normal human dermal fibroblasts. International Wound Journal. 2016; **13**(1):101-109

[83] Mei L, Xu Z, Shi Y, Lin C, Jiao S, Zhang L, et al. Multivalent and synergistic chitosan oligosaccharide-Ag nanocomposites for therapy of bacterial infection. Scientific Reports. 2020;**10**(1): 1-9

[84] Fani S, Kamalidehghan B, Lo KM, Nigjeh SE, Keong YS, Dehghan F, et al. Anticancer activity of a monobenzyltin complex C1 against MDA-MB-231 cells through induction of apoptosis and inhibition of breast cancer stem cells. Scientific Reports. 2016;**6**(1):1-15

[85] Verano-Braga T, Miethling-Graff R, Wojdyla K, Rogowska-Wrzesinska A, Brewer JR, Erdmann H, et al. Insights into the cellular response triggered by silver nanoparticles using quantitative proteomics. ACS Nano. 2014;**8**(3): 2161-2175

[86] Nadeem M, Khan R, Afridi K, Nadhman A, Ullah S, Faisal S, et al. Green synthesis of cerium oxide nanoparticles (CeO2 NPs) and their antimicrobial applications: A review. International Journal of Nanomedicine. 2020;**15**:5951

[87] Hembram KC, Kumar R, Kandha L, Parhi PK, Kundu CN, Bindhani BK. Therapeutic prospective of plant-induced silver nanoparticles: Application as antimicrobial and anticancer agent. Artificial Cells, Nanomedicine, and Biotechnology. 2018;**46**(supp. 3):S38-S51

[88] Yin IX, Zhang J, Zhao IS, Mei ML, Li Q, Chu CH. The antibacterial mechanism of silver nanoparticles and its application in dentistry. International Journal of Nanomedicine. 2020;**15**:2555

[89] Khatoon N, Alam H, Khan A, Raza K, Sardar M. Ampicillin silver nanoformulations against multidrug resistant bacteria. Scientific Reports. 2019;**9**(1):1-10

[90] Durán N, Durán M, De Jesus MB, Seabra AB, Fávaro WJ, Nakazato G. Silver nanoparticles: A new view on mechanistic aspects on antimicrobial activity. Nanomedicine: Nanotechnology, Biology and Medicine. 2016;**12**(3):789-799

[91] Du J, Tang J, Xu S, Ge J, Dong Y, Li H, et al. A review on silver nanoparticles-induced ecotoxicity and the underlying toxicity mechanisms. Regulatory Toxicology and Pharmacology. 2018;**98**:231-239

[92] Rasheed T, Bilal M, Iqbal HM, Li C. Green biosynthesis of silver nanoparticles using leaves extract of Artemisia vulgaris and their potential

biomedical applications. Colloids and Surfaces B: Biointerfaces. 2017;**158**: 408-415

[93] Baghani M, Es-haghi A. Characterization of silver nanoparticles biosynthesized using Amaranthus cruentus. Bioinspired, Biomimetic and Nanobiomaterials. 2020;**9**(3):129-136

[94] Keshari AK, Srivastava R, Singh P, Yadav VB, Nath G. Antioxidant and antibacterial activity of silver nanoparticles synthesized by Cestrum nocturnum. Journal of Ayurveda and Integrative Medicine. 2020;**11**(1):37-44

[95] Ahmad A, Wei Y, Syed F, Tahir K, Rehman AU, Khan A, et al. The effects of bacteria-nanoparticles interface on the antibacterial activity of green synthesized silver nanoparticles. Microbial Pathogenesis. 2017;**102**: 133-142

[96] Salayová A, Bedlovičová Z, Daneu N, Baláž M, Lukáčová Bujňáková Z, Balážová Ľ, et al. Green synthesis of silver nanoparticles with antibacterial activity using various medicinal plant extracts: Morphology and antibacterial efficacy. Nanomaterials. 2021;**11**(4):1005

[97] Awwad AM, Salem NM, Abdeen AO. Green synthesis of silver nanoparticles using carob leaf extract and its antibacterial activity. International Journal of Industrial Chemistry. 2013; **4**(1):1-6

[98] Bhakya S, Muthukrishnan S, Sukumaran M, Muthukumar M. Biogenic synthesis of silver nanoparticles and their antioxidant and antibacterial activity. Applied Nanoscience. 2016; **6**(5):755-766

[99] Ramteke C, Chakrabarti T, Sarangi BK, Pandey R-A. Synthesis of silver nanoparticles from the aqueous extract of leaves of Ocimum sanctum for enhanced antibacterial activity. Journal of Chemistry. 2013;**2013**

[100] Krishnaraj C, Jagan E, Rajasekar S, Selvakumar P, Kalaichelvan P, Mohan N. Synthesis of silver nanoparticles using Acalypha indica leaf extracts and its antibacterial activity against water borne pathogens. Colloids and Surfaces B: Biointerfaces. 2010;**76**(1):50-56

[101] Hungund BS, Dhulappanavar GR, Ayachit NH. Comparative evaluation of antibacterial activity of silver nanoparticles biosynthesized using fruit juices. Journal of Nanomedicine & Nanotechnology. 2015;**6**(2):1

[102] Velmurugan P, Anbalagan K, Manosathyadevan M, Lee K-J, Cho M, Lee S-M, et al. Green synthesis of silver and gold nanoparticles using Zingiber officinale root extract and antibacterial activity of silver nanoparticles against food pathogens. Bioprocess and Biosystems Engineering. 2014;**37**(10): 1935-1943

[103] Dhand V, Soumya L, Bharadwaj S, Chakra S, Bhatt D, Sreedhar B. Green synthesis of silver nanoparticles using Coffea arabica seed extract and its antibacterial activity. Materials Science and Engineering: C. 2016;**58**:36-43

[104] Salem W, Haridy M, Sayed W, Hassan N. Antibacterial activity of silver nanoparticles synthesized from latex and leaf extract of Ficus sycomorus. Industrial Crops and Products. 2014;**62**: 228-234

[105] Sheikholeslami S, Mousavi SE, Ashtiani HRA, Doust SRH, Rezayat SM. Antibacterial activity of silver nanoparticles and their combination with zataria multiflora essential oil and

methanol extract. Jundishapur Journal of Microbiology. 2016;**9**(10)

[106] Sk I, Khan MA, Haque A, Ghosh S, Roy D, Homechuadhuri S, et al. Synthesis of gold and silver nanoparticles using Malva verticillata leaves extract: Study of gold nanoparticles catalysed reduction of nitro-Schiff bases and antibacterial activities of silver nanoparticles. Current Research in Green and Sustainable Chemistry. 2020;**3**:100006

[107] Sun Q, Cai X, Li J, Zheng M, Chen Z, Yu C-P. Green synthesis of silver nanoparticles using tea leaf extract and evaluation of their stability and antibacterial activity. Colloids and Surfaces A: Physicochemical and Engineering Aspects. 2014;**444**:226-231

[108] Antony JJ, Sivalingam P, Siva D, Kamalakkannan S, Anbarasu K, Sukirtha R, et al. Comparative evaluation of antibacterial activity of silver nanoparticles synthesized using Rhizophora apiculata and glucose. Colloids and Surfaces B: Biointerfaces. 2011;**88**(1):134-140

[109] Lee J-H, Lim J-M, Velmurugan P, Park Y-J, Park Y-J, Bang K-S, et al. Photobiologic-mediated fabrication of silver nanoparticles with antibacterial activity. Journal of Photochemistry and Photobiology B: Biology. 2016;**7**(162): 93-99

[110] Murugan K, Senthilkumar B, Senbagam D, Al-Sohaibani S. Biosynthesis of silver nanoparticles using acacia leucophloea extract and their antibacterial activity. International Journal of Nanomedicine. 2014;**6**(9): 24-31

[111] Khalil MM, Ismail EH, El-Baghdady KZ, Mohamed D. Green synthesis of silver nanoparticles using olive leaf extract and its antibacterial activity. Arabian Journal of Chemistry. 2014;**7**(6): 1131-1139

[112] Kaman P, Dutta P, Bhattacharyya A. Synthesis of Gold Nanoparticles from Metarhizium Anisopliae for Management of Blast Disease of Rice and its Effect on Soil Biological Index and Physicochemical Properties. 2022. DOI: 10.21203/rs.3.rs-2080559/v1

[113] Ahmad A, Siddique JA, Laskar MA, Kumar R, Mohd-Setapar SH, Khatoon A, et al. New generation Amberlite XAD resin for the removal of metal ions: A review. Journal of Environmental Sciences. 2015;**31**:104-123

[114] Huang W, Wang C, Duan H, Bi Y, Wu D, Du J, et al. Synergistic antifungal effect of biosynthesized silver nanoparticles combined with fungicides. International Journal of Agriculture and Biology. 2018;**20**(5):1225-1229

[115] Akintelu SA, Olugbeko SC, Folorunso AS. Green synthesis, characterization, and antifungal activity of synthesized silver nanoparticles (AgNPS) from Garcinia Kola pulp extract. BioNanoScience. 2022;**12**(1): 105-115

[116] Rizwana H, Alzahrani T, Alwahibi MS, Aljowaie RM, Aldehaish HA, Alsaggabi NS, et al . Phytofabrication of silver nanoparticles and their potent antifungal activity against Phytopathogenic fungi. PRO. 2022;**10** (12):25-58

[117] Oliveira SS, Braga GC, Cordeiro NK, Stangarlin JR, Alves HJ. Green synthesis of silver nanoparticles with Euphorbia tirucalli extract and its protection against microbial decay of strawberries during storage. Journal of Food Science and Technology. 2022; **59**(5):2025-2034

[118] Purohit A, SR harma, RS hiv Ramakrishnan, S Sharma, A Kumar, et6al. Biogenic synthesis of silver nanoparticles (AgNPs) using aqueous leaf extract of Buchanania lanzan Spreng and evaluation of their antifungal activity against Phytopathogenic fungi. Bioinorganic Chemistry and Applications. 2022;**6**(44):5-8

[119] Vinodhini S, Vithiya BSM, Prasad TAA. Green synthesis of silver nanoparticles by employing the Allium fistulosum, Tabernaemontana divaricate and Basella alba leaf extracts for antimicrobial applications. Journal of King Saud University-Science. 2022; **34**(4):101939

[120] Malik MA, Batterjee MG, Kamli MR, Alzahrani KA, Danish EY, Nabi A. Polyphenol-capped biogenic synthesis of noble metallic silver nanoparticles for antifungal activity against Candida auri. Journal of Fungi. 2022;**8**(6):639

[121] Dutta T, Chowdhury SK, Ghosh NN, Chattopadhyay AP, Das M, Mandal V. Green synthesis of antimicrobial silver nanoparticles using fruit extract of Glycosmis pentaphylla and its theoretical explanations. Journal of Molecular Structure. 2022;**1247**: 131-361

[122] Ghetas HA, Abdel-Razek N, Shakweer MS, Abotaleb MM, Paray BA, Ali S, et al. Antimicrobial activity of chemically and biologically synthesized silver nanoparticles against some fish pathogens. Saudi Journal of Biological Sciences. 2022;**29**(3): 1298-1305

[123] Mishra S, Kumavat S. Green synthesis, characterization and antimicrobial activity of silver nanoparticles using Uraria picta leaves extract. Micro and Nanosystems. 2022; **14**(3):212-225

[124] Nallappan D et al. Green biosynthesis, antioxidant, antibacterial, and anticancer activities of silver nanoparticles of Luffa acutangula leaf extract. BioMed Research International. 2021

[125] Attallah NG, Elekhnawy E, Negm WA, Hussein IA, Mokhtar FA, Al-Fakhrany OM. In vivo and in vitro antimicrobial activity of biogenic silver nanoparticles against Staphylococcus aureus clinical isolates. Pharmaceuticals. 2022;**15**(2):194

[126] Khan M, Karuppiah P, Alkhathlan HZ, Kuniyil M, Khan M, Adil SF, et al. Green synthesis of silver nanoparticles using Juniperus procera extract: Their characterization, and biological activity. Crystals. 2022;**12**(3): 420

[127] Win TT, Khan S, Bo B, Zada S, Fu P. Green synthesis and characterization of Fe3O4 nanoparticles using chlorella-K01 extract for potential enhancement of plant growth stimulating and antifungal activity. Scientific Reports. 2021;**11**(1): 1-11

[128] Thanighaiarassu RR, Nambikkairaj B, Ramya DR. "Green synthesis of silver nanoparticles and characterization using plant leaf essential oil compound citral and their antifungal activity against human pathogenic fungi". Journal of Pharmacognosy and Phytochemistry. 2018;7(6):902-907

[129] Sanguiñedo P, Estevez MB, Faccio R, Alborés S. Biogenic silver nanoparticles from the fungus Punctularia atropurpurascens for the control of microorganisms. Mundo nano. Revista interdisciplinaria en

nanociencias y nanotecnología. 2019; **12**(22):10-20

[130] Elbahnasawy MA, Shehabeldine AM, Khattab AM, Amin BH, Hashem AH. Green biosynthesis of silver nanoparticles using novel endophytic Rothia endophytica: Characterization and anticandidal activity. Journal of Drug Delivery Science and Technology. 2021;**62**: 102-401

[131] Azarbani F, Shiravand S. Green synthesis of silver nanoparticles by Ferulago macrocarpa flowers extract and their antibacterial, antifungal and toxic effects. Green Chemistry Letters and Reviews. 2020;**13**(1):41-49

[132] Abdallah BM, Ali EM. Green synthesis of silver nanoparticles using the lotus lalambensis aqueous leaf extract and their anti-candidal activity against oral candidiasis. ACS Omega. 2021;**6**(12):8151-8162

[133] Mortezagholi B, Movahed E, Fathi A, Soleimani M, Forutan A, Mirhosseini NZ, et al. Plant-mediated synthesis of silver-doped zinc oxide nanoparticles and evaluation of their antimicrobial activity against bacteria cause tooth decay. Microscopy Research and Technique. 2022;**85**(11):3553-3564

[134] Lee WF, Huang YC. Swelling and antibacterial properties for the superabsorbent hydrogels containing silver nanoparticles. Journal of Applied Polymer Science. 2007;**106**(3):1992-1999

[135] Wild C, Weiderpass E, Stewart BW. World Cancer Report: Cancer Research for Cancer Prevention. International Agency for Research on Cancer. IARC Press; 2020

[136] Dziedzic A, Kubina R, Bułdak RJ, Skonieczna M, Cholewa K. Silver nanoparticles exhibit the dose-dependent anti-proliferative effect against human squamous carcinoma cells attenuated in the presence of berberine. Molecules. 2016;**21**(3):365

[137] Ahn E-Y, Jin H, Park Y. Assessing the antioxidant, cytotoxic, apoptotic and wound healing properties of silver nanoparticles green-synthesized by plant extracts. Materials Science and Engineering: C. 2019;**101**:204-216

[138] Gomathi A, Rajarathinam SX, Sadiq AM, Rajeshkumar S. Anticancer activity of silver nanoparticles synthesized using aqueous fruit shell extract of Tamarindus indica on MCF-7 human breast cancer cell line. Journal of Drug Delivery Science and Technology. 2020;**55**:101376

[139] He Y, Li X, Zheng Y, Wang Z, Ma Z, Yang Q, et al. A green approach for synthesizing silver nanoparticles, and their antibacterial and cytotoxic activities. New Journal of Chemistry. 2018;**42**(4):2882-2888

Chapter 6

Silver Nanoparticles in the Cultural Heritage Conservation

Marwa Ben Chobba, Maduka L. Weththimuni, Mouna Messaoud, Clara Urzi, Ramzi Maalej and Maurizio Licchelli

Abstract

Our cultural heritage is our invaluable social and environmental resource and concern. Moreover, it is a key global economic driver. However, they are subjected to deterioration process and aging. Particularly, microorganisms are nowadays considered harmful agents of biodeterioration of artistic materials due to the fact that their interactions with the material cause not only an esthetical damage due to their visible growth on the surface, but they may affect the interested materials in different ways and at different degrees via mechanical and biochemical processes leading to the formation of pitting, scaling and, in the worst scenario, to the loss of material by its detachment. To protect our shared tangible cultural heritage from biodeterioration and preserve it for future generations, several methods have been developed. Notably, using nanomaterials, with antimicrobial features, has been considered an interesting and economical method to preserve valuable heritage materials. In this chapter, we will present an overview of the decay mechanisms that participate in the deterioration of tangible artworks, in particular microorganisms' colonization. Next, current works that have been developed to use silver nanoparticles to protect heritage items from microbial colonization and prevent their deterioration have been detailed.

Keywords: silver, nanoparticles, microorganisms' colonization, stone biodeterioration, conservation, heritage materials

1. Introduction

Cultural heritage consists of tangible or intangible assets that history has left for a country and its future generations [1]. Heritage refers to the cultural legacies that we inherit from the past, maintain in the present, and pass on to succeeding generations. Moreover, investments in cultural heritage are frequently considered as having positive effects on a local economy, including increased employment and revenue as well as cultural consumption. However, famous works of art and historical artifacts have been marred. The alteration of cultural assets is the consequence of the continuous cycle of disintegration, by losing cohesion or strength, and reconstruction. Generally, in a natural condition, matter undergoes processes of alteration, degradation, or decomposition, which means that original chemical, physical, and optical characteristics are lost [2].

Among the decay mechanisms, the most known are as follows: (i) Physical or mechanical processes: where the behavior of the material is modified without altering the chemical composition of the material due to the contribution of several mechanical forces such as traction and compression. (ii) Chemical processes: where the matter is transformed due to a chemical reaction. (iii) Biological processes: where living organisms, such as microorganisms, insects, plants, rodents, and others, can chemically attack the material [1]. For that reason, the preservation of monuments becomes an obligation. Several products and methodologies have been studied and developed for the preservation of monuments from decay [3].

Traditionally, solvents (both aromatic and non-aromatic), chelating agents in addition to other cleaning products (including strong and mild acids and bases), have been widely used in the restoration field and are applied on different materials such as stones, paintings, and wall paintings [4]. However, the irreversibility of this technique and the risk of altering the artwork as well as the toxicity of certain products make this method not suitable for application on historic buildings [4]. Biocleaning has been suggested and studied as an alternative to the traditional cleaning techniques [5]. This method is based on using microorganisms for the treatment of deteriorated historical materials by removing undesired sulfates, nitrates, and organic matter [6]. Despite the efficiency of this method under laboratory conditions, the need of a specified condition to ensure the viability of these microorganisms makes its practical application very difficult and subject to further study [4]. The laser has been considered a friendly technique for cleaning heritage structures as well as for the environment. Its ability to alter and affect the artifacts' surface on a variety of materials, such as wood [7], paper [8], stone [9], easel paintings [10], and wall paintings [11] makes this technique attract high attention. Nevertheless, the high cost of this method is still the barrier to its widespread use.

Several products have been also developed and used as protective and consolidation agents with the aim to preserve the historic buildings from decay such as polymers [12, 13], biopolymers [14], gels [15], microemulsions [16], and ionic liquids [17]. However, the application of the abovementioned products requires caution owing to the toxicological risks to the artwork and the operator in addition to the high cost of maintenance. More recently, nanomaterials have been proposed as an alternative method in the maintenance of the historical artifacts, particularly in the consolidation and protection treatments of damaged art materials [18]. Unlike conventional materials such as polymers that are frequently used in conservation, engineered nanomaterials do not modify the original physical and chemical properties of artifacts as well as have a low environmental impact [19].

When the dimensions of particles decrease to be about 1–100 nanometers, the properties of materials change considerably from those at larger scales. In this sense, nanomaterials have larger surface areas compared with similar masses of larger-scale materials, which enhance their chemical reactivity [18]. Moreover, due to the particle size, these nanomaterials can deeply penetrate into the damaged stone materials.

The application of Nanoscience in the field of artifacts as a preservative agent reverts to the end of the 1980s through the restoration of the wall paintings (i.e., Renaissance paintings) in the "*Brancacci Chapel*" in Florence, Italy [20]. After that, different kinds of inorganic nanomaterials have been used as consolidant products [21] such as calcium hydroxide ($Ca(OH)_2$) nanoparticles, which have been applied on rocks (limestones and dolostones) and mortars [22–24]. In addition, consolidant products based on $Mg(OH)_2$ nanoparticles have been used for the conservation of cultural heritage [25] and paper-based cultural archives [26]. Barium hydroxide

($Ba(OH)_2$) has been extensively used as a consolidant product for carbonate stones for decades [27, 28]. Strontium hydroxide ($Sr(OH)_2$) nanoparticles obtained by bottom–up approach and their use as a consolidant product and also as a desulfating agent for stone, mortars, and wall paintings were proposed [22, 29]. More recently, studies concerning the behavior of commercial water-based silica dispersions, with different particles size (from 9 to 55 nm), as consolidant products on heritage materials have been performed on limestone (Lecce stone) [30]. Silicon-based hybrid polymer nanocomposites have also taken an interest in the domain of preservation of historical materials. Alkoxysilane-based formulations, notably, tetraethoxysilane (TEOS) and methyltrimethoxysilane (MTMOS), have been the most broadly used stone consolidants principally due to their ability to penetrate easily inside the porous matrix [18]. Another approach has been developed in order to reduce the cracking of the consolidant products through the addition of different metal colloidal oxide particles to the TEOS-based polymeric resins [31]. Recently, Verganelaki et al. proposed a novel crack-free calcium oxalate-silica nanocomposite for protecting historic building materials [32]. During the last years, the use of polymeric materials with hydrophobic properties has been broadly studied. The incorporation of inorganic oxide nanoparticles, such as aluminum oxide (Al_2O_3), silicon dioxide (SiO_2), and tin dioxide (SnO_2), has been performed as a method to improve the hydrophobic character of synthetic polymers such as silicone or hybrid siloxanes polymers, which leads to super-hydrophobic coatings with higher water-repellent capacity. Recently, the area of interest dealing with the treatment of monument materials for their preservation from microorganisms' colonization and pollution is shifting into the application of nanotechnology with self-cleaning properties [18]. Heterogeneous photocatalysis using semiconductors such as TiO_2 and ZnO has been considered an interesting method, used extensively to preserve heritage materials [33–39]. The particular activity of these oxides, based on the generation of reactive oxygen species (ROS) when they are exposed to UV light, makes them very attractive compounds [40].

Recently, nanoparticles of different elements (e.g., Ag, Ti, Cd, Fe, Pd, Zn, Pt, and Co) have been investigated due to their interesting attributes. Indeed, the low cost, nontoxicity, chemical stability, biological inertness, suitability toward visible or near-UV light, sustained photoactivity, high conversion efficiency as well as higher quantum yield [41]. In fact, these nanoparticles could react with an extensive range of substrates and have high adaptability to several environments and good absorption in the domain of the solar spectrum [41]. In addition, treatments by using the above-mentioned nanoparticles can provide water-repellent features and inhibit the damage generation performed by water [42]. In particular, Ag NPs display good antimicrobial properties in comparison to other metals, which makes them widely used for several applications, in particular, conservation of heritage materials from biodeterioration [43]. To our knowledge, this is the first report on the application of silver nanoparticles in cultural heritage conservation.

Several variables need to be taken into account in order to have an adequate protective coating by selecting materials and procedures to be used. The intrinsic properties of historic material, the environmental factors, the compatibility between tested products and affected heritage materials, the durability of applied products, the biological colonization, in addition to the application method, are significant factors that have to be taken into consideration [22]. However, it is essential to identify firstly the main factors considered as threats to heritage materials, in order to delay the deterioration processes [1] as much as possible.

2. Decay mechanisms of cultural heritage materials

Historic materials are subject to several decay phenomena. The main deteriorating factors will be explicated in the following sections.

2.1 Water

Many processes that can alter historical materials are often driven by water. In fact, water is one of the most significant abiotic factors of deterioration of porous stones, which are more susceptible to water. Highly porous materials with average porosity values between 30 and 40% and above have been widely used over the world as historic building materials [33]. When water penetrates the pores through capillary force, water performs its deteriorating effect via both chemical and physical phenomena such as dissolution of carbonate component of the stone, freezing/thawing cycles, salt crystallization, and deposition [44, 45]. Such processes can affect the physical as well as mechanical properties of stone in addition to inducing some changes in its structure such as modification of porosity and pore structure, loss of stone cohesion, and development of cracks [46]. The need of the surface protection, especially against water penetration, is mandatory to conserve historical artifacts.

Dissolution is usually caused by water that frequently contains acid due to dissolved carbon dioxide. It is able to dissolve minerals from a rock body, leaving cavities in the rock. These cavities may generate cave features [47].

The freezing-thawing cycles of the water inside the stone pores result in the development of internal stresses, which can lead to cracking and progressive desegregation of material [46]. Particularly, this phenomenon is widely observed in cold regions exposed to excessive freezing and thawing during the year. Tensions that frequently increase and decrease due to the formation of successive layers of ice result in the fracture of the material [46]. The formation of ice in the pores creates a crystallization pressure, that is, a pressure caused by the crystallization of ice against the pores in walls of the material. This phenomenon can affect even stone with very fine pores, (diameter smaller than 0.1 μm). Water freezes at temperatures considerably under $0^{\circ}C$; the freezing of water causes internal stresses that are the direct cause of the damage. The pressure rises as the temperature decreases [48].

2.2 Salt crystallization

Salt crystallization affects heritage materials constructed from rocks. Indeed, plenty of architectural heritage materials around the world are constructed principally of carbonate rocks. However, historical monuments are frequently affected by several degradation and alteration processes typically due to salt crystallization [49]. Such phenomenon is considered as one of the most powerful weathering factors in the case of porous materials, particularly limestone rocks. This problem affects monuments and historical buildings around the world, from the cooler and wetter conditions found in the United Kingdom to hyper-arid desert environments and Mediterranean climates. Macroscopically, such a type of deterioration process produces enormously aggressive damages such as loss of material, erosion, flaking, exfoliation, and occasionally, even the complete disaggregation of the material [44]. Once a salt-rich solution enters into the pore structure of stone material, in favorable supersaturating and thermodynamic conditions, the salt crystallization occurs. Besides, nucleation and growth of salt crystals inside the pore spaces can promote alteration and degradation of the stone [50]. Many factors would affect the damage of the material such as the supersaturation of the salt, the shape of crystals, the pore size, and the repulsion force between the salt and the walls of the pores [51]. Porosity represents a critical parameter in the process of salt crystallization since it regulates the fluid mobility inside the material. Salt crystals exert pressure on the capillary wall, namely the linear pressure, which is proportional to crystal size and salt concentration and inversely proportional to the radius of the pore.

2.3 Microorganisms' colonization

Physical and chemical processes were commonly assumed to be the principal factors in the degradation of materials. However, since 1967 and in later decades, dogma has changed, and today, it is believed that one of the main causes of deterioration of archaeological stone materials is the microbial action, leading to biodeterioration [1]. Biodeterioration can be defined as "any undesirable change in a material brought about by the vital activities of organisms." The biodeterioration of cultural heritage can also be defined as "the physical or chemical damage caused by microorganisms on objects, monuments, or buildings that belong to the cultural heritage" [1]. The main microorganisms that play a role in biodeterioration are both autotrophic and heterotrophic bacteria, fungi, cyanobacteria, algae, and lichens [1]. Microbial populations present in a stone foundation are generally the result of successive colonization of different microorganisms over many years [52].

Many types of research on the biological deterioration of stone monuments have been performed, and results showed that bacteria, algae, fungi, and lichens are the microorganisms mainly responsible for artifacts biodegradation [53]. Whereas, mosses have received relatively less attention because their impact has been considered mainly esthetic [53]. In fact, bacteria were isolated from many archaeological statues (e.g., in the Museum in Zagreb, Croatia [54]), deteriorated marble (e.g., Moscow Kremlin masonry [55]), from wall paintings [56] and found in caves and catacombs [57]. Fungi were also isolated frequently from historic limestone buildings and antique marble from different sites in many countries such as Germany, Portugal, Italy, Russia, Namibia, and Spain. On the other hand, it was stated that cyanobacteria and chlorophyta (green algae) are considered the pioneering inhabitants in the colonization of stone works of art [58].

Microorganisms damage stone in different ways, including discoloration, water retention, material breakage, growth stimulation of heterotrophic organisms, disintegration of the material, degradation (corrosion), formation of patinas, and alkaline dissolution. On the other hand, many antique and precious historical items are suffering from serious microbial invasions that cause in some cases the closure of archeological sites, as in the case of the caves of Lascaux in France [59].

Biofilms, which are made by a complex microbial organization mainly containing bacteria and fungi, contribute also to degrade historical statues. All they need are surface, moisture, nutrients, and microorganisms. The role of biofilms in the biodeterioration of cultural heritage materials has been reported for several decades and is related to: (a) accumulation of water that enters the matrix causing swelling and amplified conductivity; (b) alterations in pH values and ionic concentrations; (c) releasing enzymes that lead to embrittlement and loss of mechanical stability and others [1, 60].

There is a general agreement that microbial activity contributes to altering and deteriorating historical relics. The question now is why and how microorganisms are able to degrade the works of art. Some artifacts such as parchment, wood, leather, or textiles are made with organic materials, which microbes like to feast on. It used to be common for artists to use egg tempera, pigments mixed with egg yolks as a binding agent, which has left their paintings more susceptible to infestation. Fungi are among the most active microorganisms in these processes where they can use organic support as nutrients. Moreover, in the case of inorganic supports, several metabolites that are excreted may react with the support in different ways [53]. In the case of stone materials, their mineralogical nature, surface features, and environmental conditions play a role in the colonization by microorganisms [1]. Due to their heterotrophic nature, fungi promote biological deterioration by synthesizing certain compounds, such as inorganic and organic acids, which allow them to utilize the contents of the inorganic supports [53]. On the other hand, inorganic materials have been also considered as good substrates for a high number of different microorganisms. It was reported that most kinds of microorganisms are able to attack and degrade marble materials [53]. It is now generally agreed that fungi and bacteria not only cause serious aesthetical problems but also reside and penetrate inside the materials, resulting in a material loss due to acid corrosion, enzymatic degradation, and mechanical attack [61]. Acidic compounds can also corrode metals. Algae and cyanobacteria, as an example, are able to adapt to different light qualities [61] and environments (humid, semiarid, and arid) [62] and to be developed on stone in archaeological hypogea (e.g., an underground chamber). Such an environment is characterized by low light intensities (i.e., crypts, caves, and catacombs). Cyanobacteria cause esthetic damage to marble surfaces; their endolithic mode of life contributes to the collapse of rock crystalline structures [62].

2.4 Climatic conditions

The climatic conditions play an important role in deterioration mechanisms. Particularly, the architectural structures and historical monuments, which are exposed outdoors, are susceptible to several climatic conditions such as the wind that wears the rock eroding it, the solar radiation causing discoloration, the rain, the humidity, and the snow that induce physical and chemical wear [1]. These factors affect significantly the stability of stone matrix by inducing chemical corrosion through hydration and oxidation reactions, the dissolution of carbonates as well as solubilization of some mineral components.

On the other hand, the climate has an effect on microbial colonization species. In moderate or humid climates, the fungal communities on the rock are predominant. Whereas in arid and semiarid environments, such as those found in the Mediterranean area, the climatic conditions are overly extreme, consequently the communities shift toward the so-called black yeasts and microcolonial fungi [61]. Moreover, the role of bacteria in the weathering of rock depends mostly on the environmental conditions: while bacteria might evolve in humid environments and form biofilms inside the porous space of building stone, in arid and semiarid environments, their occurrence might be limited [61].

Climate change is another potential risk since it increases the anticipated degradation rates and/or participates to the occurrence of new decay phenomena [63]. In fact, climatic changes are able to aggravate the chemical, physical, and biological processes producing deterioration by influencing the composition and/or structure of the affected heritage materials [64]. Climate change could also affect the intensity as well as the frequency of dangerous events such as floods, landslides, and droughts with unavoidable extensive impacts, also on the cultural heritage materials. The United Nations Educational, Scientific and Cultural Organization (UNESCO) has considered that changes in the temperature, wind intensity, precipitation, desertification, atmospheric moisture in addition to the interaction between air pollution and climatic changes have been identified as threats to cultural heritage [65].

There are synergisms that take place between different parameters, leading to direct and indirect, immediate, and long-term impacts on a widespread variety of objects, buildings, materials, and sites. Changes in the conservation conditions because of climate-related degradation processes are inevitable phenomena for immovable and movable cultural heritage materials [63].

3. Silver nanoparticles

Due to their distinctive physical and chemical features, silver nanoparticles (Ag NPs) have been extensively used in a several domains, including medicine, health care, food, and others [66]. The exceptional features of silver NPs are the main factor for the extensive range of their application in several fields. **Figure 1** illustrates the main application fields of Ag NPs.

In general, Ag NPs with size lower than 100 nm contain around 20–15,000 silver atoms and have exceptional chemical, physical, and biological features compared with their bulk materials. The optical, catalytic, and thermal characteristics of Ag NPs are intensely affected by their size and shape. In particular, Ag NPs are characterized by the localized surface plasmon resonance (LSPR), which has attracted considerable attention over the past few years in the photocatalytic field. Thanks to their broad-spectrum antimicrobial capacity, Ag NPs have also become the highest used sterilizing nanomaterials. It is well known that silver compounds are highly effective to kill microorganisms such as bacteria in addition to certain fungi and viruses. The use of silver as a biocide in the form of nanoparticles has grown significantly due to the higher fraction of surface atoms of silver nanoparticles, which leads to a better antimicrobial activity compared with bulk silver metal. Numerous studies have been reported in order to clarify the inhibitory effect of silver nanoparticles on bacteria. Reports declared that the electrostatic attraction between negatively charged

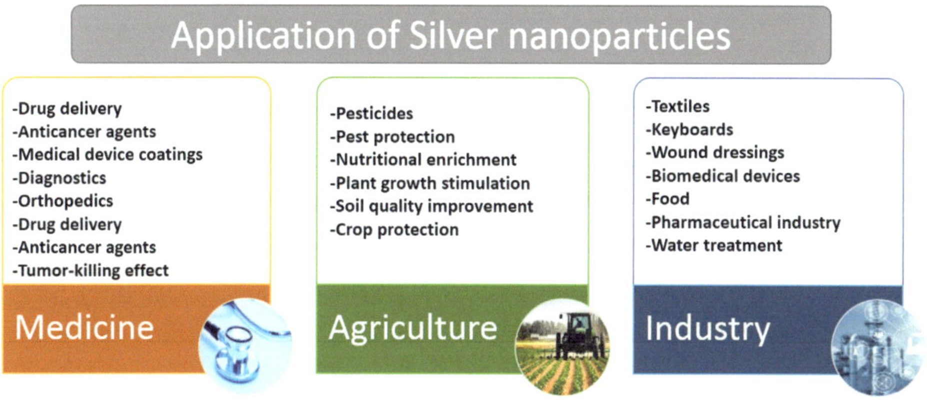

Figure 1.
Silver nanoparticles' application in different domains [66–68].

bacterial cells and the positively charged nanoparticles plays a key role in the activity of nanoparticles as bactericidal materials. Nevertheless, Sondi and Salopek-Sondi synthesized negatively charged silver NPs [69]; and results revealed that they show excellent antibacterial activity against *E. coli*. Authors suggested that the negative surface charge of nanoparticles interacts with the building elements of the bacterial membrane, provoking structural changes in addition to degradation and, lastly, cell death. EDS analysis proved the existence of the elementary silver in the membranes of treated bacteria; such observation confirmed the incorporation of silver NPs into the membrane structure. Yamanaka et al. [70] proposed that the inhibitory activity of silver on the microorganisms growth could be due to the toxic effect on the DNA replication causing the inactivation of vital cellular proteins. Another suggestion reported by Choi and Hu [71], based on the fact that the cell death can be attributed to the formation of free radicals, particularly, reactive oxygen species (ROS) from the surface of Ag. The generation of free radicals can attack membrane lipids and lead to a collapse of membrane function. It has also been stated that Ag ions can bind with the sulfhydryl groups of the proteins disturbing the membrane function that bound enzymes of the respiratory chain [72]. Although the biocidal effect of nanosized silver particles against bacteria was widely studied, the mechanism of the growth-inhibitory effects of Ag nanoparticles on microorganisms is still not fully understood. The different hypotheses concerning the antibacterial effect of silver nanoparticles are described in detail in the literature [73].

4. Application of silver nanoparticles for the preservation of heritages materials

Heritage buildings are subject to different decay phenomena. In particular, microorganisms' colonization affects them in different ways and at different degrees starting from esthetical problems and formation of patinas, often causing total breakage and loss of the materials. Archival documents and library collections such as paper, parchment, textile, glue, leather, and photographs are of high cultural importance. However, microorganisms could easily attack them, particularly in high humidity conditions. Therefore, protecting heritage materials has become a necessity. Applying silver NPs has been suggested as a promising and effective method. In this context, several groups have worked on exploiting silver nanoparticles to protect heritage items from microbial colonization and prevent their deterioration.

Gutarowska et al. are among the first groups who worked on applying silver nanoparticles to museum collections and archives, which are contaminated by microorganisms [74]. The authors worked precisely on analyzing the sensitivity of microorganisms to Ag NPs in six different museums and archives in Poland. Results showed that microorganisms can be effectively removed from the surface of artifacts at a concentration of 90 parts per million (ppm). While a concentration of only 45 ppm was able to remove 94% of all tested microorganisms with the exception of *Staphylococcus xylosus* and *Bacillus subtilis* that showed higher resistance. The study revealed that silver NPs could be considered a promising disinfectant for the surfaces of historical materials and archival documents. The Lodz team developed also a novel procedure for disinfecting cultural artifacts by nebulizing Ag NPs from a dispersion over the paper, textiles, or canvas [75]. These researchers discovered that the degree of relative humidity affected the misting disinfection procedure, as moisture facilitates the entry of NPs into microbial walls. Furthermore, they found that vegetative cells (mycelium and bacteria) are more susceptible to Ag NPs antimicrobial activity than fungal or bacterial spores. Next, F. Bellissima et al. have synthesized silver nanoparticles according to a seed-based procedure in order to preserve Serena stone (SS, 5–10% open porosity) from biodeterioration [76]. The ability of Ag NPs to hinder bacterial colonization was investigated through the inactivation of *Bacillus subtilis*. Aiming to chemically graft silver NPs to SS surface, tetraethylorthosilicate (TEOS) was used as a grafting agent, and dimethylamine was used as a catalyst to promote the condensation of a silane precursor on NPs surface. Antimicrobial activity on stones was evaluated through "spot on spot" method. Results showed that applied nano-coatings did not induce significant color changes on the stone surface after treatment ($\Delta E < 2$) and provided good antibacterial activity with a reduction in cell viability between 50 and 80%, with the most efficient nanoparticles concentration equal to 6.7 $\mu g/cm^2$. A.M.M. Essa and M.K. Khallaf have studied the efficiency of Ag NPs suspensions as an antimicrobial agent to preserve Edfu (Sandstone) and the tomb of Teti's son Teti-ankh-km at Sakkara (Limestone) [77]. Both archaeological buildings are localized in Egypt. Silver NPs were elaborated biologically through volatile metabolites formed during the aerobic growth of *N. halobius* and then mixed with acrylic polymers (Primal AC33 polymer) and silicon as a consolidation polymer at a concentration of 40 mg/ml. Small Ag particles with diameters of 10–20 nm were obtained as displayed by scanning electron microscopy analysis. The antimicrobial properties of prepared NPs were studied against *Streptomyces parvulus* as Gram-positive bacterial strain and *Aspergillus niger* as a fungal strain. Findings showed high antibacterial performance of the applied nano-coatings against *S. parvulus*. The treated sandstone samples showed a significant decrease in the percentage of bacterial cell recovery (98.4% and 97.2%) by using functionalized silicon polymer and acrylic polymer, respectively. Moreover, treated limestone specimens exhibited a clear reduction in the percentage of *S. parvulus* cell recovery that achieved 97.1% and 98.6% with the impregnated acrylic polymer and the functionalized silicon polymer, respectively. On the other hand, a significant inhibition growth of *A. niger* on both the surfaces of the sandstone and limestone samples coated by silver NPs and silicon or acrylic polymers occurred compared with untreated stones on which a prominent growth of *A. niger* was observed. MacMullen et al. have investigated that the efficiency of silver NPs (<100 nm) enhanced aqueous silane/siloxane emulsions performance to protect and improve the facade interface of historical buildings [78]. Different emulsions were prepared and applied on mortar samples. Results showed that all treated samples exhibited water-repellent character with water contact angles higher than 100°; however, treatment induced a color modification higher than that which is considered tolerable for application on cultural heritage with $\Delta E > 5$. Authors declared also that adequate concentrations (<0.5% wt) were needed to reach considerably beneficial enhancements. Capillary absorption analysis showed that water absorption was considerably reduced in the case of treated mortars compared with the untreated ones, despite that negligible difference was observed when Ag NPs were incorporated. Therefore, the authors considered that silver treatments are not suitable for marine or flood applications where there is prolonged water exposure. Next, R. Carrillo-González et al. have prepared silver NPs through a green process using leaf aqueous extracts of *T. stans* and *F. vulgare* [79]. The inhibitory growth of biosynthesized NPs at different doses was tested against microorganisms, which are isolated from biofilms developed on the surfaces of three kinds of stony historic monument walls located in the pre-Hispanic city, Teotihuacan (Mexico). *In vitro* biocontrol of isolated microorganisms revealed that Ag NPs prepared from *F. vulgare* were more efficient to inhibit microbial growth than those prepared from *T. stans*. Results also showed that bacterial strains were less sensitive to silver NPs than fungal strains and that sensitivity is principally related to the microbial strain and the plant extract utilized to elaborate silver NPs. Therefore, the authors considered that using Ag NPs as a corrective or preventive treatment to reduce microbial colonization from historical walls was effective. K. Pietrzak et al. [80] have investigated the efficiency of Ag NPs misting as a decontamination process compared with the effect of two other disinfection procedures such as thyme essential oil microatmosphere (TEO) and low-temperature plasma (LTP), to inhibit microbial growth from two archival books through culture-dependent method and RNA analysis. In this study, two books with observable signs of biodegradation obtained from Jozef Pilsudski Regional and Municipal Public Library in Lodz (Poland) and National Archive in Prague (Czech Republic) were used. in the study. Results revealed that Ag NPs misting process was more efficient for bacterial inhibition (R = 60–100%), while the two other methods showed less effectiveness with LTP (R = 25–100%) and TEO (R = 12–100%). Furthermore, it was stated that all tested methods showed less efficiency against fungi (R = 0–99.8%). Another study performed by K. Pietrzak et al. [81] to explore Ag NPs misting method in order to disinfect historical textile materials, precisely, five pre-Columbian fibers (1250–1450 A.D., Argentina). Microscopic analyses indicated that tested items were fabricated from sisal, cotton, and wool, and they were contaminated by dust and mineral impurities. The reduction in microbial population ranged from 30.8 to 99.9%, depending on the variety of microbial colonization and its concentration. In fact, the sensitivity of microorganisms toward silver NPs was fluctuated in most resistant endospore-forming bacteria *Bacillus*, while *Oceanobacillus*, *Paracoccus, Kocuria*, and molds *Penicillium*, *Cladosporium* were more easily inhibited. Interestingly, it was found that Ag NPs misting process does not harmfully affect the pH and the chemistry of textiles. The same group worked on evaluating the anti-biofilm capabilities of Ag NPs to protect textiles from *Pseudomonas sp* [82]. In the study, textile materials were collected during excavations in Santa Rosa de Tastil, Puna Argentina (1967–1969). Microscopic observations revealed that the bacterial strain including *Pseudomonas aeruginosa* and *Clostridium sp.* was presented on examined archaeological textile items through lipolytic and proteolytic activities. Results demonstrated that using Ag NPs with particles size ranged from 10 to 80 nm and a concentration of 90 ppm was effective to protect archaeological textiles against *P. aeruginosa* growing by 63%–97%. Authors stated that the inhibition ability of NPs was influenced by the kind of strain and the exposition time. Aerial algae are a central biological factor contributing to the degradation of building materials as well as facades. In this regard, P. Nowicka-Krawczyk et al. have examined the effect of silver NPs on *Apatococcus lobatus* as an algal model due to its frequent presence in aerial biofilms formed on historical building facades [83]. Authors stated that changes in the chloroplasts structure and the photosynthetic activity of the tested cells have been observed through confocal laser microscopy and digital image analysis due to exposition to Ag NPs. The rate of growth inhibition was estimated through a biomass test, and results showed that the average biomass of control samples was 3.7 mg/l chl *a*. While after exposition to Ag NPs, the biomass was decreased by 26% for 8 ppm, 56% for 15 ppm, 65% and 68% for 20 and 107 ppm, respectively. Authors recommended that Ag NPs could be employed as a biocide against aerial algal coatings, with caution in terms of nanoparticle concentration. More recently, Z. Li et al. [84] have developed a colorimetric sensor array (CSA) by using printed inks of 10 nm Ag NPs with different capping agents as an alternative process to passive sampling indicators, which were traditionally used by conservators. CSAs have been used as "optoelectronic nose" to detect and identify individual compounds and highly similar complex mixtures such as oxidants, acids, and aldehydes from ppm to tens of ppb levels. In fact, for artifacts, the levels of pollutant exposure are generally recommended at a few ppb or even sub-ppb concentrations. Authors found that the developed CAS was ultrasensitive and able to identify quantitatively 11 common contaminants related to the preservation of cultural heritage items and museums. Thanks to changes in localized surface plasmon resonance of metallic NPs, in particular silver, due to sintering of solid-state NPs, alterations in color during exposure to pollutants can be observed.

Nanocomposite coatings have been suggested as heterojunction systems to enhance the light absorption of semiconductors such as (TiO_2 and ZnO) and separate the charge carrier by localizing photogenerated holes and electrons to different regions in the system. In this regard, Ag NPs have been suggested to be coupled with semiconductors to develop several biocidal coatings with other interesting features to be applied to heritage building materials in order to conserve them. In this context, M. Aflori et al. [85] have compared the antibacterial/antifungal efficacy of silsesquioxane-based hybrid nanocomposites with methacrylate units containing only Ag NPs to its counterpart containing Ag/TiO_2 NPs as protective coatings for heritage stone surfaces. The antibacterial and antifungal efficiency of the prepared nanocomposites was evaluated through the inactivation of *Escherichia coli* (*E. coli*) and *Candida albicans* (*C. albicans*), respectively. Findings revealed that the coatings composed of Ag/TiO_2 NPs showed higher activity against *E. coli* and *C. albicans* than the coatings composed of only Ag NPs. Pinho et al. prepared hydrophobic and self-cleaning Ag-TiO2-SiO2 coatings for outdoor applications [86]. Findings revealed that prepared nanocomposites efficiently inhibit the penetration of water into the pores of the tested stones based on the total water uptake values, which were close to zero for all treated samples, and remarkably attained lower values compared with untreated ones. Such results prove the water-repellent feature of the developed coating since water causes many decay mechanisms to porous materials. Indeed, when water penetrates the pores by capillary force, it performs its deteriorating effect through salt crystallization; chemical dissolution of the carbonate component of the stone, and freezing/thawing cycles. On the other hand, incorporation of high amounts of Ag (5% w/w) leads to an undesired color alteration on studied stones and would prevent the sol–gel transition of the nanocomposite coatings (10% w/w). In fact, one of the conditions to accept the application of such treatment, according to Normal 20/85 [87], is that a protective product should not cause a visible alteration and must be stable over time. Moreover, authors found that the incorporation of silver into TiO_2-SiO_2 network notably improves the photodegradation activity of the coating containing 1% (w/v) TiO_2 thanks to the enhanced absorption under visible light and larger surface area of the photocatalyst. L. Graziani et al. [88] have compared the efficiency of Ag NPs incorporated in TiO_2 sol nanocoatings for the preservation of brick-based heritage materials compared with their Cu-TiO_2 counterpart. The concentration of Ag or Cu nanoparticles was set at 1% (molar weight/TiO_2 weight), and the total color variation at that concentration was higher than the value accepted for treatments on historical buildings, which is fixed to 5. The self-cleaning test showed that untreated samples achieved an efficacy of around 9%, while specimens treated with Cu/or Ag-TiO_2 were about 46 and 33%, respectively. Interestingly, Ag-TiO_2 nanocomposite showed a higher ability to hinder algal growth than Cu-TiO_2 treatment, which revealed the efficiency of Ag NPs as a biocide agent. J. Becerra et al. [89] investigated

Nanomaterials	Particles size	Substrate	Obtained results	References
SiO_2/Ag NPs	5–15 nm	Buildings Materials stone	Biocidal efficiency up to > 90%. Hierarchical roughness attributed to creation of Ag/SiO_2 NPs clusters. Improved contact with the cell walls. Easy removal of the dead cells. Good durability of the treatment.	[90]
Ag NPs	10 nm	Paper: Ancient manuscripts and books.	Suitable sensors for the detection, identification, and quantification of pollutants to the preservation of cultural heritage objects.	[84]
Ag-TiO_2NPs	11.8 ± 3.6 nm	Fossiliferous limestone	Treatment with self- cleaning activity. Ag NPs prevent the growth of *E. coli* and *S. cerevisiae* in the dark. Ameliorated inhibition growth of the coatings applied on the stone up to a 20% (*S. cerevisiae*) and 70% (*E. coli*).	[93]
PMAA@Ag NPs	200 nm	Paper, wood, or stone	PMAA@Ag nanostructures revealed an efficient fungicidal activity against *Aspergillus niger*. Nanomaterials can be used on cultural heritage materials for preventive conservation	[94]
Ag/N-SiO_2NPs	50–100 nm	Cement mortars	Efficient anti-fouling surface treatments. Multifunctional superhydrophobic/ biocide treatment was obtained.	[91]
Ag NPs	7.2 nm ± 1.8 nm	Paper	A total removal of dirt from the paper samples, Softening of the dirt from the canvas, without affecting the integrity or leaving residues of the treated art works.	[95]

Table 1.
Application of silver nanomaterials for the treatment of historical artifacts (references are presented in a chronological order).

the suitability and the efficacy of nanocomposite treatments based on TiO_2 and/or Ag NPs to inhibit the biodegradation of limestone heritage materials. Findings revealed that Ag and TiO_2 nanocomposites when stabilized by citrate reach a good biocide effect by inducing a substantial reduction of the biopatina growth: and retaining the color modifications at an acceptable level. Next, R. Zarzuela et al. [90] have developed multifunctional treatment by incorporating Ag NPs grafted to functionalized silicon dioxide NPs in an organically modified silica matrix (Ag/N-SiO_2). This treatment was applied to different kinds of stone specimens that are frequently used to construct heritage building by spray method in order to enhance the mechanical resistance of the stone-based monuments, improve water permeability, and produce biocidal surfaces. Authors stated that thanks to coupling Ag NPs with SiO_2, the biocidal efficiency reached 90% values, since Ag NPs raise the stability of the treatment and improve the contact with the cell walls. Furthermore, the synergistic effect facilitates the removal of dead cells, ameliorating treatment durability. The same Ag/N-SiO_2 nanocomposite was used in another study performed by M. Domínguez et al. [91] as an anti-fouling agent. Authors studied the impact of surface features such as roughness/texture, surface free energy, and charge of the surface on the biocidal efficiency of NPs and the interaction with the cell walls. Authors stated that the functionalization of SiO_2 NPs with the positively charged –NHx groups considerably improved their interaction capability with the negatively charged cell walls through the electrostatic forces. For that reason, this interaction between biocide agent and *Phormidium sp*. was the lowest due to the presence of a positively charged mucopolysaccharide sheath cell wall. R. Zarzuela et al. [92] have elaborated Ag/modified-TiO_2 nanoparticles (Ag /N-TiO_2) for numerous applications (environmental, self-cleaning, antifouling, etc.). The biocidal efficiencies of Ag/TiO_2 nano-powders were evaluated through mixed cultures composed of three biofilm-forming phototrophic microorganisms isolated from building façades. Whereas, the photocatalytic activity of the obtained nanocomposite was assessed through the degradation of methylene blue dye under solar spectrum lamp. Results showed that the incorporation of Ag NPs into TiO_2 surface noticeably ameliorates the absorption of the semiconductor under the visible spectrum thanks to the surface plasmon resonance band provided by silver ions and, consequently, improved photodegradation capacity of TiO_2 NPs. Moreover, findings showed that Ag/TiO_2 was able to hinder microalgae growth under visible irradiation, whereas pure TiO_2 NPs were ineffective. The prepared functionalized nanocomposites were then used as protective treatment for porous ceramic materials. Nanocoatings exhibited superhydrophobic performance, high capacity to degrade pollutants, and good biocidal properties on surface stones [93]. **Table 1** summarizes the main application of Ag NPs as preventive and consolidant treatments in the recent years (2019–2022).

5. Drawbacks, challenges, and perspectives

Silver nanoparticles become more broadly used in the field of protection of cultural heritage materials. However, the transport of NPs in different environments is still not clearly understood. In general, nanoparticles have the potential to leach into the environment and cause ecotoxicity in soil, water, and related biota [96]. The major form of the toxic action and toxic effects are usually related to the dissolution of the metal into metal ions, and then it is introduced into the environment. Zhang et al. [97] pointed out the necessity of investigating NPs' toxicity in natural waterways. To clearly understand the toxicity of nanoparticles in natural water, authors studied the harmfulness of Ag NPs toward unicellular green alga "*Chlorella pyrenoidosa*" in four freshwater bodies. Findings revealed that water chemistry had a deep impact on dissolution, aggregation, and algal toxicity of nanoparticles. Authors stated that the ecotoxicity of silver NPs was generally attributed to the release of the harmful ions. Indeed, the dissolved Ag^+ might be principally responsible for the toxicity of Ag NPs. On the other hand, Franco-Castillo [98] stated that Ag NPs are regularly cytotoxic and consequently have an impact on the health, the safety, and environmental implications for restorers-conservators, curators, and general public.

The potential of using engineered nanoparticles for the preservation of heritage structures has been developed for the restoration of deteriorated materials, production of self-cleaning surfaces or developing surfaces as a biocide to minimize biodeterioration. However, several studies need ideally to be carried out *in situ*, or in the real environment, such as outdoors, in archives, where the anthropogenic circumstances could be investigated appropriately. The period of the research projects, the time required for the implementation of medium-to-long-term *in situ* investigation by collaborative programs including research centers, museums, and professionals in conservation and restoration, needs to be taken into account [98]. Studies should not be limited only to the efficiency of antimicrobial treatment but also to the entire procedure involving the design, application, investigation of the progressive evolution of the treatment, detachment, leaching, or ions release. Moreover, a protocol concerning the removal or clearance of treatment needs to be taken into consideration, in accordance with conservation rules. Several works have been performed to conserve heritage materials through the application of Ag NPs, these NPs are still far away from being routine treatments. There exists limited understanding concerning the impact of engineered nanoparticles on the environment after being released from the treated surfaces. Nevertheless, it is important to control their impact on the different nearby non-target organisms and ecological processes [99].

6. Conclusion

Tackling the deterioration of cultural heritage needs a global effort. Material scientists are asked to elaborate novel nanomaterials and develop new methods for the conservation of heritage artwork. In this study, we have particularly focused on the different works that have been developed to use Ag NPs to preserve heritage materials from deterioration. In this chapter, the different degradation mechanisms of cultural heritage materials have been presented in detail. An overview about different nanomaterials used as protective and consolidation agents with the aim to preserve the historic buildings from deterioration has been provided. Next, the different works that have been performed to exploit silver nanoparticles used alone or coupled with semiconductors for cultural heritage conservation have been described. Studies showed that silver NPs revealed high capability to protect heritage materials from biodeterioration. The main risks of using silver NPs for conservation purpose in addition to the different challenges to make Ag NPs widely used in this field have also been discussed. More importantly, the effect of NPs after their application on heritage materials surfaces needs extensive studies, to be assured about them being harmless to the environment and human beings.

Author details

Marwa Ben Chobba[1*], Maduka L. Weththimuni[2], Mouna Messaoud[1], Clara Urzi[3], Ramzi Maalej[4] and Maurizio Licchelli[2]

1 Laboratory of Advanced Materials, National School of Engineering, University of Sfax, Sfax, Tunisia

2 Department of Chemistry, University of Pavia, Pavia, Italy

3 Department of Chemical, Biological, Pharmaceutical and Environmental Sciences, University of Messina, Messina, Italy

4 Faculty of Sciences of Sfax, Laboratory of Dielectric and Photonic Materials, Sfax University, Sfax, Tunisia

*Address all correspondence to: marwa.benchobba@enis.tn

References

[1] Rivera LEC, Ramos AP, Sánchez JIC, Serrano MED. Origin and control strategies of biofilms in the cultural heritage. In: Kırmusaoğlu S, editor. Antimicrobials, Antibiotic Resistance, Antibiofilm Strategies and Activity Methods. London, UK, Rijeka: IntechOpen; 2018

[2] De Leo F, Marchetta A, Capillo G, Germanà A, Primerano P, Schiavo SL, et al. Surface active ionic liquids based coatings as subaerial anti-biofilms for stone built cultural heritage. Coatings. 2021;**11**:26. DOI: 10.3390/coatings11010026

[3] Chobba MB, Weththimuni ML, Messaoud M, Bouaziz J, Licchelli M. Enhanced Gd doped TiO_2 NPs-PDMS nanocomposites as protective coatings for bio-calcarenite stone: Preliminarily analysis. In: Walha L, Jarraya A, Djemal F, Chouchane M, Aifaoui N, Chaari F, Abdennadher M, Benamara A, Haddar M, editors. Design and Modeling of Mechanical Systems - V. Cham: Springer International Publishing; 2023. pp. 885-893. DOI: 10.1007/978-3-031-14615-2_99

[4] Balliana E, Ricci G, Pesce C, Zendri E. Assessing the value of green conservation forcultural heritage: Positive and critical aspects ofalready available methodologies. International Journal of Conservation Science. 2015;7:185-202

[5] Sanmartín P, Cappitelli F, Mitchell R. Current methods of graffiti removal: A review. Construction and Building Materials. 2014;**71**:363-374

[6] Alfano G, Lustrato G, Belli C, Zanardini E, Cappitelli F, Mello E, et al. The bioremoval of nitrate and sulfate alterations on artistic stonework: The case-study of Matera Cathedral after six years from the treatment. International Biodeterioration and Biodegradation. 2011;**65**:1004-1011

[7] Tang QH, Zhou D, Wang YL, Liu GF. Laser cleaning of sulfide scale on compressor impeller blade. Applied Surface Science. 2015;**355**:334-340. DOI: 10.1016/j.apsusc.2015.07.128

[8] Kolar J, Strlič M, Müller-Hess D, Gruber A, Troschke K, Pentzien S, et al. Laser cleaning of paper using Nd:YAG laser running at 532 nm. Journal of Cultural Heritage. 2003;**4**:185-187. DOI: 10.1016/S1296-2074(02)01196-2

[9] Sabatini G, Giamello M, Pini R, Siano S, Salimbeni R. Laser cleaning methodologies for stone façades and monuments: Laboratory analyses on lithotypes of Siena architecture. Journal of Cultural Heritage. 2000;**1**:S9-S19. DOI: 10.1016/S1296-2074(00)00144-8

[10] Staicu A, Apostol I, Pascu A, Urzica I, Pascu ML, Damian V. Minimal invasive control of paintings cleaning by LIBS. Optics and Laser Technology. 2016;**77**:187-192. DOI: 10.1016/j.optlastec.2015.09.010

[11] Gaetani C, Santamaria U. The laser cleaning of wall paintings. Journal of Cultural Heritage. 2000;**1**:S199-S207. DOI: 10.1016/S1296-2074(00)00137-0

[12] Licchelli M, Malagodi M, Weththimuni ML, Zanchi C. Water-repellent properties of fluoroelastomers on a very porous stone: Effect of the application procedure. Progress in Organic Coating. 2013;**76**:495-503. DOI: 10.1016/j.porgcoat.2012.11.005

[13] Weththimuni M, Crivelli F, Galimberti C, Malagodi M, Licchelli M.

Evaluation of commercial consolidating agents on very porous biocalcarenite. International Journal of Conservation Science. 2020;**11**:251-260

[14] Ocak Y, Sofuoglu A, Tihminlioglu F, Böke H. Protection of marble surfaces by using biodegradable polymers as coating agent. Progress in Organic Coating. 2009;**66**:213-220. DOI: 10.1016/j.porgcoat.2009.07.007

[15] Domingues JAL, Bonelli N, Giorgi R, Fratini E, Gorel F, Baglioni P. Innovative hydrogels based on semi-interpenetrating p(HEMA)/PVP networks for the cleaning of water-sensitive cultural heritage artifacts. Langmuir. 2013;**29**:2746-2755. DOI: 10.1021/la3048664

[16] Baglioni P, Giorgi R, Dei L. Soft condensed matter for the conservation of cultural heritage. Comptes Rendus Chimie. 2009;**12**:61-69. DOI: 10.1016/j.crci.2008.05.017

[17] Greer AJ, Jacquemin J, Hardacre C. Industrial Applications of Ionic Liquids. Molecules. 2020;**25**:5207. DOI: 10.3390/molecules25215207

[18] Sierra-Fernandez A, Gomez-Villalba L, Rabanal M, Fort R. New nanomaterials for applications in conservation and restoration of stony materials: A review. Materiales de Construcción. 2017;**67**:107. DOI: 10.3989/mc.2017.07616

[19] Baglioni P, Carretti E, Chelazzi D. Nanomaterials in art conservation. Nature Nanotechnology. 2015;**10**:287-290. DOI: 10.1038/nnano.2015.38

[20] Baglioni P, Berti D, Bonini M, Carretti E, Dei L, Fratini E, et al. Micelle, microemulsions, and gels for the conservation of cultural heritage. Advances in Colloid and Interface Science. 2014;**205**:361-371. DOI: 10.1016/j.cis.2013.09.008

[21] Ion R-M, Doncea S-M, Caruțiu DȚ. Nanotechnologies in cultural heritage - Materials and instruments for diagnosis and treatment. In: Kyzas GZ, Mitropoulos AC, editors. Novel Nanomaterials. London, UK, Rijeka: IntechOpen; 2017. DOI: 10.5772/intechopen.71950

[22] Licchelli M, Malagodi M, Weththimuni M, Zanchi C. Nanoparticles for conservation of bio-calcarenite stone. Applied Physics A. 2014;**114**:673-683. DOI: 10.1007/s00339-013-7973-z

[23] Poggi G, Toccafondi N, Melita LN, Knowles JC, Bozec L, Giorgi R, et al. Calcium hydroxide nanoparticles for the conservation of cultural heritage: New formulations for the deacidification of cellulose-based artifacts. Applied Physics A. 2014;**114**:685-693. DOI: 10.1007/s00339-013-8172-7

[24] López-Arce P, Gómez-Villalba LS, Martínez-Ramírez S, de Buergo MÁ, Fort R. Influence of relative humidity on the carbonation of calcium hydroxide nanoparticles and the formation of calcium carbonate polymorphs. Powder Technology. 2011;**205**:263-269. DOI: 10.1016/j.powtec.2010.09.026

[25] Baglioni P, Giorgi R. Soft and hard nanomaterials for restoration and conservation of cultural heritage. Soft Matter. 2006;**2**:293-303. DOI: 10.1039/B516442G

[26] Saoud K, Saeed S, Soubaih R, Samara A, Ibala I, Ladki D, et al. Application of $Mg(OH)_2$ nanosheets for conservation and restoration of precious documents and cultural archives. Bioresources. 2018;**13**:3259-3274. DOI: 10.15376/biores.13.2.3259.3274

[27] Karatasios I, Kilikoglou V, Colston B, Theoulakis P, Watt D. Setting process of lime-based conservation mortars with barium hydroxide. Cement and Concrete Research. 2007;**37**:886-893. DOI: 10.1016/j.cemconres.2007.03.007

[28] Rodrigues JD, Pinto APF. Laboratory and onsite study of barium hydroxide as a consolidant for high porosity limestones. Journal of Cultural Heritage. 2016;**19**:467-476. DOI: 10.1016/j.culher.2015.10.002

[29] Ciliberto E, Condorelli GG, La Delfa S, Viscuso E. Nanoparticles of Sr(OH)2: Synthesis in homogeneous phase at low temperature and application for cultural heritage artefacts. Applied Physics A. 2008;**92**:137-141. DOI: 10.1007/s00339-008-4464-8

[30] Aggarwal P, Singh RP, Aggarwal Y. Use of nano-silica in cement based materials—A review. Cogent Engineering. 2015;**2**:1078018 https://doi.org/10.1080/23311916.2015.1078018

[31] Miliani C, Velo-Simpson ML, Scherer GW. Particle-modified consolidants: A study on the effect of particles on sol–gel properties and consolidation effectiveness. Journal of Cultural Heritage. 2007;**8**:1-6. DOI: 10.1016/j.culher.2006.10.002

[32] Verganelaki A, Kilikoglou V, Karatasios I, Maravelaki-Kalaitzaki P. A biomimetic approach to strengthen and protect construction materials with a novel calcium-oxalate–silica nanocomposite. Construction and Building Materials. 2014;**62**:8-17. DOI: 10.1016/j.conbuildmat.2014.01.079

[33] Chobba MB, Weththimuni ML, Messaoud M, Urzi C, Bouaziz J, Leo FD, et al. Ag-TiO2/PDMS nanocomposite protective coatings: Synthesis, characterization, and use as a self-cleaning and antimicrobial agent. Progress in Organic Coating. 2021;**158**:106342. DOI: 10.1016/j.porgcoat.2021.106342

[34] Chobba MB, Weththimuni ML, Messaoud M, Sacchi D, Bouaziz J, De Leo F, et al. Multifunctional and durable coatings for stone protection based on Gd-doped nanocomposites. Sustainability. 2021;**13**:11033. DOI: 10.3390/su131911033

[35] Weththimuni M, Ben Chobba M, Tredici I, Licchelli M. ZrO2-doped ZnO-PDMS nanocomposites as protective coatings for the stone materials. ACTA IMEKO. 2022;**11**:5. DOI: 10.21014/acta_imeko.v11i1.1078

[36] Weththimuni ML, Chobba MB, Sacchi D, Messaoud M, Licchelli M. Durable polymer coatings: A comparative study of PDMS-based nanocomposites as protective coatings for stone materials. Chemistry. 2022;**4**:60-76. DOI: 10.3390/chemistry4010006

[37] Cinteză LO, Tănase MA. Multifunctional ZnO nanoparticle: Based coatings for cultural heritage preventive conservation. In: Ares AE, editor. Thin Films. London, UK, Rijeka: IntechOpen; 2020. DOI: 10.5772/intechopen.94070

[38] Luna M, Gatica JM, Vidal H, Mosquera MJ. Use of Au/N-TiO2/SiO2 photocatalysts in building materials with NO depolluting activity. Journal of Cleaner Production. 2020;**243**:118633. DOI: 10.1016/j.jclepro.2019.118633

[39] Kapridaki C, Xynidis N, Vazgiouraki E, Kallithrakas-Kontos N, Maravelaki-Kalaitzaki P. Characterization of photoactive Fe-TiO_2 lime coatings for building protection: The role of iron content. Materials. 2019;**12**:1847. DOI: 10.3390/ma12111847

[40] Ben Chobba M, Messaoud M, Bouaziz J, De Leo F, Urzì C. The effect of heat treatment on photocatalytic performance and antibacterial activity of TiO_2 nanoparticles prepared by sol-gel method. In: Chaari F, Barkallah M, Bouguecha A, Zouari B, Khabou MT, Kchaou M, Haddar M, editors. Advances in Materials, Mechanics and Manufacturing. Cham: Springer International Publishing; 2020. pp. 71-79. DOI: 10.1007/978-3-030-24247-3_9

[41] Aziz AA, Cheng CK, Ibrahim S, Matheswaran M, Saravanan P. Visible light improved, photocatalytic activity of magnetically separable titania nanocomposite. Chemical Engineering Journal. 2012;**183**:349-356. DOI: 10.1016/j.cej.2012.01.006

[42] Colangiuli D, Calia A, Bianco N. Novel multifunctional coatings with photocatalytic and hydrophobic properties for the preservation of the stone building heritage. Construction and Building Materials. 2015;**93**:189-196. DOI: 10.1016/j.conbuildmat.2015.05.100

[43] Ben Chobba M, Weththimuni ML, Messaoud M, Bouaziz J, Salhi R, De Leo F, et al. Silver-doped TiO_2-PDMS nanocomposite as a possible coating for the preservation of serena stone: Searching for optimal application conditions. Heritage. 2022;**5**:3411-3426. DOI: 10.3390/heritage5040175

[44] La Russa MF, Ruffolo SA, Belfiore CM, Aloise P, Randazzo L, Rovella N, et al. Study of the effects of salt crystallization on degradation of limestone rocks. Periodico Di Mineralogia. 2013;**82**:113-127. DOI: 10.2451/2013PM0007

[45] Ricca M, Le Pera E, Licchelli M, Macchia A, Malagodi M, Randazzo L, et al. The CRATI Project: New Insights on the Consolidation of Salt Weathered Stone and the Case Study of San Domenico Church in Cosenza (South Calabria, Italy). Coatings. 2019;**9**:330. DOI: 10.3390/coatings9050330

[46] Martins L, Vasconcelos G, Lourenco P, Palha C. Influence of the freeze-thaw cycles on the physical and mechanical properties of granites. Journal of Materials in Civil Engineering. 2015;**28**:04015201. DOI: 10.1061/(ASCE)MT.1943-5533.0001488

[47] Abd El-Aal A. Climate change and its impact on monumental and historical buildings towards conservation and documentation Ammon temple, Siwa Oasis, Egypt. Journal of Earth Science & Climatic Change. 2016;**7**. DOI: 10.4172/2157-7617.1000339

[48] Stryszewska T, Kańka S. Forms of damage of bricks subjected to cyclic freezing and thawing in actual conditions. Materials. 2019;**12**:1165. DOI: 10.3390/ma12071165

[49] Belfiore CM, La Russa MF, Pezzino A, Campani E, Casoli A. The Baroque monuments of Modica (Eastern Sicily): Assessment of causes of chromatic alteration of stone building materials. Applied Physics A. 2010;**100**:835-844. DOI: 10.1007/s00339-010-5659-3

[50] Weththimuni ML, Licchelli M, Malagodi M, Rovella N, Russa ML. Consolidation of bio-calcarenite stone by treatment based on diammonium hydrogenphosphate and calcium hydroxide nanoparticles. Measurement. 2018;**127**:396-405. DOI: 10.1016/j.measurement.2018.06.007

[51] Scherer GW. Stress from crystallization of salt. Cement and Concrete Research. 2004;**34**:1613-1624. DOI: 10.1016/j.cemconres.2003.12.034

[52] De Leo F, Marchetta A, Urzì C. Black fungi on stone-built heritage: Current knowledge and future outlook. Applied Sciences. 2022;**12**:3969. DOI: 10.3390/app12083969

[53] Abdelhafez AAM, El-Wekeel FM, Ramadan EM, Abed-Allah AA. Microbial deterioration of archaeological marble: Identification and treatment. Annals of Agricultural Science. 2012;**57**:137-144. DOI: 10.1016/j.aoas.2012.08.007

[54] Čavka M, Glasnović A, Janković I, Sikanjic P, Perić B, Brkljacic B, et al. Microbiological analysis of a Mummy from the archeological Museum in Zagreb. Collegium Antropologicum. 2010;**34**:803-805

[55] Doronina N, Li T, Ivanova E, Trotsenko I. Methylophaga murata sp. nov.: A Haloalkaliphilic aerobic methylotroph from deteriorating marble. Mikrobiologiia. 2005;**74**:511-519

[56] Pangallo D, Kraková L, Chovanová K, Šimonovičová A, De Leo F, Urzì C. Analysis and comparison of the microflora isolated from fresco surface and from surrounding air environment through molecular and biodegradative assays. World Journal of Microbiology and Biotechnology. 2012;**28**:2015-2027. DOI: 10.1007/s11274-012-1004-7

[57] De Leo F, Iero A, Zammit G, Urzì C. Chemoorganotrophic bacteria isolated from biodeteriorated surfaces in cave and catacombs. International Journal of Speleology. 2012;**41**:1-12. DOI: 10.5038/1827-806X.41.2.1

[58] Macedo M, Miller A, Dionísio A, Saiz-Jimenez C. Biodiversity of cyanobacteria and green algae on monuments in the Mediterranean Basin: An overview. Microbiology (Reading, England). 2009;**155**:3476-3490

[59] Bastian F, Alabouvette C. Lights and shadows on the conservation of a rock art cave: The case of Lascaux Cave. International Journal of Speleology. 2009;**38**

[60] Milanesi C, Baldi F, Borin S, Vignani R, Ciampolini F, Faleri C, et al. Biodeterioration of a fresco by biofilm forming bacteria. International Biodeterioration and Biodegradation. 2006;**57**:168-173. DOI: 10.1016/j.ibiod.2006.02.005

[61] Sterflinger K, Piñar G. Microbial deterioration of cultural heritage and works of art — tilting at windmills? Applied Microbiology and Biotechnology. 2013;**97**:9637-9646. DOI: 10.1007/s00253-013-5283-1

[62] Lamprinou V, Mammali M, Katsifas EA, Pantazidou AI, Karagouni AD. Phenotypic and molecular biological characterization of cyanobacteria from marble surfaces of treated and untreated sites of propylaea (Acropolis, Athens). Geomicrobiology Journal. 2013;**30**:371-378. DOI: 10.1080/01490451.2012.690021

[63] Bertolin C. Preservation of cultural heritage and resources threatened by climate change. Geosciences. 2019;**9**:250. DOI: 10.3390/geosciences9060250

[64] Sesana E, Gagnon A, Ciantelli C, Cassar J, Hughes J. Climate change impacts on cultural heritage: A literature review. WIREs Climate Change. 2021;**12**. DOI: 10.1002/wcc.710

[65] UNESCO World Heritage Centre. Climate Change and World Heritage. Report on Predicting and Managing the Impacts of Climate Change on World Heritage and Strategy to Assist States Parties to Implement Appropriate Management Responses. Paris, France: UNESCO; 2007

[66] Zhang X-F, Liu Z-G, Shen W, Gurunathan S. Silver nanoparticles: Synthesis, characterization, properties, applications, and therapeutic approaches. International Journal of Molecular Sciences. 2016;**17**:1534. DOI: 10.3390/ijms17091534

[67] Kale SK, Parishwad GV, Patil AS. Emerging agriculture applications of silver nanoparticles. ES Food & Agroforestry. 2021;**3**:17-22

[68] Jaskulski D, Jaskulska I, Majewska J, Radziemska M, Bilgin A, Brtnicky M. Silver nanoparticles (AgNPs) in urea solution in laboratory tests and field experiments with crops and vegetables. Materials. 2022;**15**:870. DOI: 10.3390/ma15030870

[69] Sondi I, Salopek-Sondi B. Silver nanoparticles as antimicrobial agent: A case study on E. coli as a model for Gram-negative bacteria. Journal of Colloid and Interface Science. 2004;**275**:177-182. DOI: 10.1016/j.jcis.2004.02.012

[70] Yamanaka M, Hara K, Kudo J. Bactericidal actions of a silver ion solution on *Escherichia coli*, studied by energy-filtering transmission electron microscopy and proteomic analysis. Applied and Environmental Microbiology. 2005;**71**:7589-7593. DOI: 10.1128/AEM.71.11.7589-7593.2005

[71] Choi O, Hu Z. Size dependent and reactive oxygen species related nanosilver toxicity to nitrifying bacteria. Environmental Science & Technology. 2008;**42**:4583-4588. DOI: 10.1021/es703238h

[72] Rai M, Yadav A, Gade A. Silver nanoparticles as a new generation of antimicrobials. Biotechnology Advances. 2009;**27**:76-83. DOI: 10.1016/j.biotechadv.2008.09.002

[73] Bruna T, Maldonado-Bravo F, Jara P, Caro N. Silver nanoparticles and their antibacterial applications. International Journal of Molecular Sciences. 2021;**22**:7202. DOI 10.3390/ijms22137202

[74] Gutarowska B, Skora J, Zduniak K, Rembisz D. Analysis of the sensitivity of microorganisms contaminating museums and archives to silver nanoparticles. International Biodeterioration and Biodegradation. 2012;**68**:7-17. DOI: 10.1016/j.ibiod.2011.12.002

[75] Gutarowska B, Rembisz D, Zduniak K, Skóra J, Szynkowska M, Gliścińska E, et al. Optimization and application of the misting method with silver nanoparticles for disinfection of the historical objects. International Biodeterioration and Biodegradation. 2012;**75**:167-175. DOI: 10.1016/j.ibiod.2012.10.002

[76] Bellissima F, Bonini M, Giorgi R, Baglioni P, Barresi G, Mastromei G, et al. Antibacterial activity of silver nanoparticles grafted on stone surface. Environmental Science and Pollution Research. 2014;**21**:13278-13286. DOI: 10.1007/s11356-013-2215-7

[77] Essa AMM, Khallaf MK. Biological nanosilver particles for the protection of archaeological stones against microbial colonization. International Biodeterioration and Biodegradation. 2014;**94**:31-37. DOI: 10.1016/j.ibiod.2014.06.015

[78] MacMullen J, Zhang Z, Dhakal HN, Radulovic J, Karabela A, Tozzi G, et al. Silver nanoparticulate enhanced aqueous silane/siloxane exterior facade emulsions and their efficacy against algae and cyanobacteria biofouling. International Biodeterioration and Biodegradation. 2014;**93**:54-62. DOI: 10.1016/j.ibiod.2014.05.009

[79] Carrillo-González R, Martínez-Gómez MA. Inhibition of microorganisms involved in deterioration of an archaeological site by silver nanoparticles produced by a green synthesis method. Science of The Total Environment. 2016;**565**:872-881

[80] Pietrzak K, Otlewska A, Danielewicz D, Dybka K, Pangallo D, Kraková L, et al. Disinfection of archival documents using thyme essential oil, silver nanoparticles misting and low temperature plasma. Journal of Cultural Heritage. 2017;**24**:69-77. DOI: 10.1016/j.culher.2016.10.011

[81] Pietrzak K, Puchalski M, Otlewska A, Wrzosek H, Guiamet P, Piotrowska M, et al. Microbial diversity of pre-Columbian archaeological textiles and the effect of silver nanoparticles misting disinfection. Journal of Cultural Heritage. 2017;**23**:138-147. DOI: 10.1016/j.culher.2016.07.007

[82] Pietrzak K, Otlewska A, Puchalski M, Gutarowska B, Patricia G. Antimicrobial properties of silver nanoparticles against biofilm formation by Pseudomonas aeruginosa on archaeological textiles. Applied Environmental Biotechnology. 2016;**1**:1. DOI: 10.18063/AEB.2016.02.001

[83] Nowicka-Krawczyk P, Zelazna-Wieczorek J, Koźlecki T. Silver nanoparticles as a control agent against facades coated by aerial algae—A model study of Apatococcus lobatus (green algae). PLoS One. 2017;**12**:e0183276. DOI: 10.1371/journal.pone.0183276

[84] Li Z, Wang Z, Khan J, LaGasse MK, Suslick KS. Ultrasensitive monitoring of museum airborne pollutants using a silver nanoparticle sensor array. ACS Sensors. 2020;**5**:2783-2791. DOI: 10.1021/acssensors.0c00583

[85] Aflori M, Simionescu B, Bordianu I-E, Sacarescu L, Varganici C-D, Doroftei F, et al. Silsesquioxane-based hybrid nanocomposites with methacrylate units containing titania and/or silver nanoparticles as antibacterial/antifungal coatings for monumental stones. Materials Science and Engineering B. 2013;**178**:1339-1346. DOI: 10.1016/j.mseb.2013.04.004

[86] Pinho L, Rojas M, Mosquera MJ. Ag–SiO_2–TiO_2 nanocomposite coatings with enhanced photoactivity for self-cleaning application on building materials. Applied Catalysis B: Environmental. 2015;**178**:144-154. DOI: 10.1016/j.apcatb.2014.10.002

[87] NORMAL 20/85. Interventi conservativi: progettazione esecuzione e valutazione preventiva, Milan, Italy. 1996

[88] Graziani L, Quagliarini E, D'Orazio M. The role of roughness and porosity on the self-cleaning and anti-biofouling efficiency of TiO2-Cu and TiO2-Ag nanocoatings applied on fired bricks. Construction and Building Materials. 2016;**129**:116-124. DOI: 10.1016/j.conbuildmat.2016.10.111

[89] Becerra J, Zaderenko AP, Sayagués MJ, Ortiz R, Ortiz P. Synergy achieved in silver-TiO2 nanocomposites for the inhibition of biofouling on limestone. Building and Environment. 2018;**141**:80-90. DOI: 10.1016/j.buildenv.2018.05.020

[90] Zarzuela R, Carbú M, Gil MLA, Cantoral J, Mosquera M. Ormosils loaded with SiO_2NPs functionalized with Ag as multifunctional superhydrophobic/biocidal/consolidant treatments for buildings conservation. Nanotechnology. 2019;**30**:345701. DOI: 10.1088/1361-6528/ab1ff0

[91] Domínguez M, Zarzuela R, Moreno-Garrido I, Carbú M,

Cantoral JM, Mosquera MJ, et al. Antifouling nano-Ag/SiO2 ormosil treatments for building materials: The role of cell-surface interactions on toxicity and bioreceptivity. Progress in Organic Coating. 2021;**153**:106120. DOI: 10.1016/j.porgcoat.2020.106120

[92] Zarzuela R, Moreno-Garrido I, Gil MLA, Mosquera MJ. Effects of surface functionalization with alkylalkoxysilanes on the structure, visible light photoactivity and biocidal performance of Ag-TiO2 nanoparticles. Powder Technology. 2021;**383**:381-395. DOI: 10.1016/j.powtec.2021.01.050

[93] Zarzuela R, Carbú M, Gil A, Cantoral J, Mosquera MJ. Incorporation of functionalized Ag-TiO2NPs to ormosil-based coatings as multifunctional biocide, superhydrophobic and photocatalytic surface treatments for porous ceramic materials. Surfaces and Interfaces. 2021;**25**:101257. DOI: 10.1016/j.surfin.2021.101257

[94] Kainourgios P, Tziveleka L-A, Kartsonakis IA, Ioannou E, Roussis V, Charitidis CA. Silver nanoparticles grown on cross-linked poly (methacrylic acid) microspheres: Synthesis, characterization, and antifungal activity evaluation. Chemosensors. 2021;**9**:152. DOI: 10.3390/chemosensors9070152

[95] Oliveira MJA, Otubo L, Pires A, Brambilla RF, Carvalho AC, Santos PS, et al. Silver nanoparticles-based hydrogels synthetized by ionizing radiation for cleaning of tangible cultural heritage surfaces. Radiation Physics and Chemistry. 2022;**199**:110345. DOI: 10.1016/j.radphyschem.2022.110345

[96] Reyes-Estebanez M, Morales O, Chan M, Granados-Echegoyen C, Camacho-Chab J, Pereañez Sacarias J, et al. Antimicrobial engineered nanoparticles in the built cultural heritage context and their ecotoxicological impact on animals and plants: A brief review. Heritage Science. 2018;**6**

[97] Zhang L, Li J, Yang K, Liu J, Lin D. Physicochemical transformation and algal toxicity of engineered nanoparticles in surface water samples. Environmental Pollution. 2016;**211**:132-140. DOI: 10.1016/j.envpol.2015.12.041

[98] Franco-Castillo I, Hierro L, de la Fuente JM, Seral-Ascaso A, Mitchell SG. Perspectives for antimicrobial nanomaterials in cultural heritage conservation. Chem. 2021;7:629-669. DOI: 10.1016/j.chempr.2021.01.006

[99] de Gannes V, Hickey, WJ. Genetic Adaptations of Bacteria for Metabolism of Polycyclic Aromatic Hydrocarbons. In: Cravo-Laureau C, Cagnon C, Lauga B, Duran R, editors. Microbial Ecotoxicology. Cham: Springer; 2017. pp 133-164. DOI: 10.1007/978-3-319-61795-4_7